JACK DARR'S SERVICE CLINIC No. Three

No. 761
$7.95

JACK DARR'S SERVICE CLINIC No. Three

By Jack Darr

TAB BOOKS
Blue Ridge Summit, Pa. 17214

FIRST EDITION

FIRST PRINTING—JUNE 1975

Printed in the United States
of America

Hardbound Edition: International Standard Book No. 0-8306-5761-4

Paperbound Edition: International Standard Book No. 0-8306-4761-9

Library of Congress Card Number: 75-1696

Foreword

Jack Darr, as Service Editor of Radio-Electronics magazine, has become something of a "landmark" in television servicing. A repairman's repairman, Jack Darr has been turning tough dogs into pussycats for as long as TV sets have been acting up. The material in this book represents the best of Darr—his penetrating way of seeing the source of the trouble while most of us are wondering over the symptoms, his plain-talk wisdom and down-to-earth humor, and his timeless philosophy (think, think, think).

Readers of Radio-Electronics will doubtless recognize some of the material included here, for the book is formed from Darr's columns in that magazine: *The Service Clinic* and *In The Shop*. We have taken his columns and the letters he's received from readers, and arranged them into a logical chapter sequence, according to general symptom category.

The treatment here tends toward the informal. That's Darr's way. Each chapter winds up with a question-and-answer session relating to material presented within that chapter. The questions, for the most part, are extracted from letters to Darr from servicemen who got stumped at some point during their trouble analysis. The answers—a great many of them eventually verified by the technicians who asked the questions initially—serve a twofold purpose: they comprise a valuable file of case histories that are bound to prove extremely helpful to any practicing serviceman who runs into a blind alley on a repair job, and they point the way to success in solving unusual problems that recur in specific models.

Contents

1 Color Problems 11

Color Oscillator: It's Easy to Know—Categorizing the Trouble—Diode Grid Leakage—Smeary Color—Convergence Controls—Comb Color Troubles Out of Your Hair!—AFPC Troubles—Reader Questions.

2 Sweep & Deflection Problems 44

Heeding The Clues—Same Trouble, Different Causes!—Vertical Output Circuits—Vertical Output Transformer—Yoke—White Vertical Line—Vertical Controls—Complementary Vertical Sweep Circuits—Reader Questions.

3 Component Peculiarities 78

Checking the Chip—A Dropping Diode—What Is It?—ICs, TV and You—Replacement Parts—Solid State "Tubes"—Getting to Know Varactor Diodes—Replacing Hard-to-Find Transistors—Reader Questions.

4 CRT Troubles: Flyback, Focus, Voltages, Etc. 118

Heater-to-Cathode Shorts—Focusing—Flyback Transformers—Reader Questions.

5 Tuners, Noise, and AGC Problems 146

Mixer Noise—TV Boosters—Interference—Reader Questions.

6 Diagnosing Trouble & Shaking Down Intermittents 168
Asking the Right Questions—The Trouble Tree—Case History—Recognizing Intermittents—Reader Questions.

7 TV Sets and CATV 188
Cable Systems—The Videotape Machine—Reader Questions.

8 Test Equipment 198
Capacitances—Cable Reaks—Homebrew AC Ammeter Wattmeter—Homebrew Scope "Prescaler"—The VOM Around the Car—Reader Questions.

9 The Open Forum 215
Voltage Problems—Horizontal and Vertical Problems—AFC Problems—Sync Problems—Non-TV Equipment

Index 243

Color Problems Chapter 1

Color problems are going to be with us for as long as we have color sets. And even though these problems will be fewer in the coming years—in the sense that they won't occur quite as often—they'll still happen. There is no amount of modularization, self-checking, built-in redundancy, or special circuitry that can completely eliminate all the problems that can occur when a crt displays the red, green, and blue in a manner not intended by the station, network, or TV manufacturer. In this chapter, we look at some of those problems that recur most frequently: intermittent color; absence of color; messy, smeary color; bad convergence; and a few others. Perhaps an ideal jumping-off point would be that place in the set where a great many of the problems actually start—the 3.57945 MHz color oscillator (which we'll refer to from here on out as the 3.58 MHz oscillator).

COLOR OSCILLATOR: IT'S EASY TO KNOW

The color oscillator is a circuit that causes a good deal of confusion. It shouldn't. It's a nice friendly circuit; not at all complicated after you get acquainted. In fact, it's exactly the same as a couple of old circuits you've been fixing for years! Learn the basic reactions, and it's easy. Not much test equipment needed, either; a color-bar generator, VTVM, and a diddle-stick!

It's a crystal-controlled oscillator, running at 3.58 MHz, which everyone calls "three-fifty-eight" for short. To reproduce colors correctly, it must be tightly locked in phase with the burst signal. This consists of little samples of the original 3.58 MHz signal from the transmitter, to which all of the color signals are "referred." All color signals are ac waveforms with a certain *phase* relationship to the *burst*. So, if the local color oscillator gets slightly out of phase with the burst, we still see colors, but they're not the right ones.

The local oscillator is held in phase by a circuit that works exactly like the familiar horizontal afc. It's called afpc: (automatic frequency and *phase* control), mostly to let you know it isn't afc. The action is the same; the incoming burst signal is compared to the local oscillator signal. If it isn't precisely in phase, a small dc control voltage is generated which pulls the local oscillator back in step.

When we were young, we were taught that a crystal-controlled oscillator was very stable. It is. However, most of us got the idea that a crystal oscillator could not be "tuned" or its frequency varied. This isn't so. By varying either capacitance or inductance in the crystal oscillator circuit, you *can* make it change frequency. Not much, but plenty for what we need.

In practical circuits, this is done with what we call a reactance tube. This is a circuit with a tube which has its plate current out of phase with its plate voltage. So, it looks like a *reactance* to any circuit in which it's used. By varying the control-grid voltage of this tube, we can change this reactance; this in turn changes the oscillator frequency. The

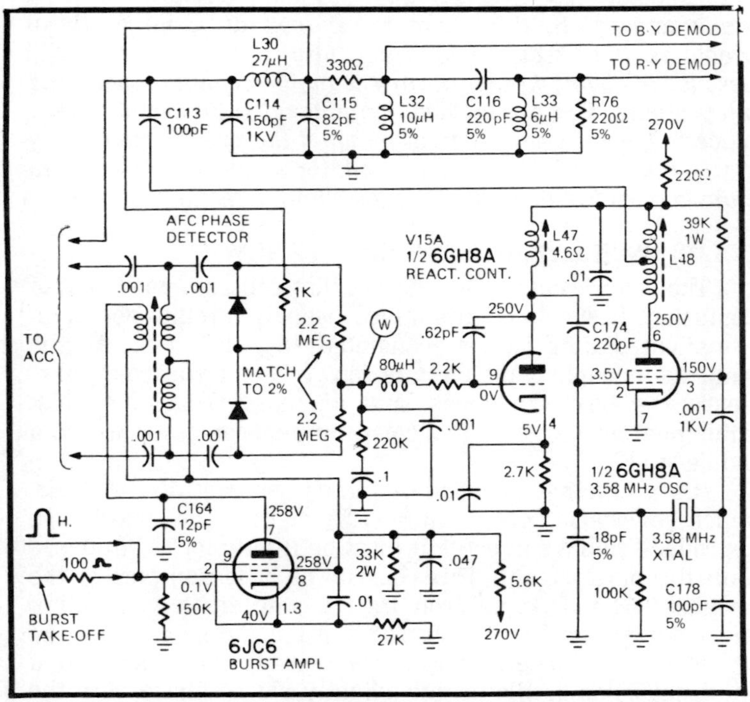

Fig. 1-1. TV crystal oscillator and phase control system, 3.58 MHz.

schematic in Fig. 1-1 shows a typical crystal oscillator circuit, with the reactance tube, burst amplifier, etc. Note the dual-diode phase detector which varies the reactance tube grid voltage.

Like the horizontal oscillator, our crystal oscillator must be able to run, without any control at all, very close to the right frequency. If it will, the frequency control can hold it with ease. So, this leads us to our first simple test. Kill the frequency control, and adjust the oscillator frequency to exactly where it must be. (Afc adjustment again!)

To do this, we need a color picture on the screen of the TV; preferably a color-bar pattern. First, we ground the input of the burst amplifier. This kills the incoming color sync. Our colors will promptly fall out of sync. If you'll notice, they will make the familiar slanting lines that we see when the horizontal oscillator isn't locked. They'll make bands of red, green, and blue, or rainbows in slanting lines across the picture. The farther the color oscillator is off, the more rainbows we see.

By tuning the reactance coil, we can make fewer and fewer rainbows, until we eventually wind up with only three vertical color bars. A careful twiddle of the adjustment will make these lock in place on the color-bar pattern. Like horizontal bars, they'll float, since there's no sync, but the fact that we can get this momentary lock shows that the oscillator is running on frequency.

Now, we take the ground off the burst amplifier and check the sync. The colors should lock in and hold. Just as with afc, if this makes the colors fall *out* of sync, we know where the trouble is. In the afpc—not in the oscillator! So we check out the burst amplifier, the transformer, and the diodes.

Oscillator Failure

If we can see strong black-and-white bars on the screen, but no colors, this is usually a symptom of oscillator failure. There is one tricky reaction here that can fool you. In many sets, if the 3.58 MHz oscillator goes out, you will see a bar pattern which is blue and green! (This seems to be due to a little bit of burst getting through; the burst is in this area.) So, you may be misled into hunting for a "loss of the reds." Always check the 3.58 MHz oscillator first, just to be sure.

Another definite check is tracing the color-bar waveform through the color-difference amplifiers to the picture-tube *grids*. If you see the familiar comb pattern at these points, but it is *flat-topped*, your oscillator is out. The color signal *is* going through the bandpass amplifiers *and* the demodulators;

without the oscillator, there is no demodulation, which would give you the normal "rocker" or lazy-S patterns on the grids.

Oscillator Testing

Testing these oscillators isn't hard. Even a narrowband scope will show you enough vertical deflection so that you can get an idea of the p-p voltage. (To find out, check an oscillator that is working; this will show you what you will see with *your* scope.) You do *not* have to see individual cycles of this waveform. This wouldn't tell you anything anyway. You check phase and frequency by simply *looking* at the color-bar pattern and noting whether the colors are there and in the right place.

Control Testing

You'll find quite a few sets with wrong colors, weak color sync, and so on. If you *do* see colors, the oscillator must be running. If they're wrong, the frequency is off. If it's off far enough, the afpc won't be able to hold it. When you change stations, the color sync will fall out. This can be due to one of two things. The oscillator circuit may be mistuned, or you may have a crystal which is slightly off frequency.

This is easy to pin down. Just run the full afpc setup adjustment. If the oscillator jumps far off frequency when the burst is killed, it could be mistuned. (Symptom: many small rainbows on screen.) Run the tuneup procedure for the oscillator. In most sets, simply adjust the reactance coil until you see only the three color bars, and a momentary color lock.

When the afpc is put back it should lock the colors tightly. In a lot of sets, the holding range can be checked by reading the dc control voltage developed by the afpc on the grid of the reactance tube. Normal voltage here, in sync, should be zero volts. The afpc should develop equal and opposite polarity voltages as the reactance coil is tuned through its full range. A good ballpark figure for this would be ±5V, zero volts with the oscillator locked.

But! What if you read about −3V or +3V with the oscillator in *sync*? Or, if the control voltage will not cross zero but locks (with weak color sync), what if the value is something like −3V? This is not too common, but it can happen. It is almost always due to a *crystal* that is just a little bit off frequency! By carefully diddling the afpc, you can literally haul it back on frequency, but your hold will be very bad. The oscillator is *pulling* against the afpc all the time, like a car with a soft front tire. For a quick check, substitute an exact replacement crystal. (Does this sound like tuning up an

FM detector to get an S-curve? Sure is. Same thing. That's the other familiar circuit I mentioned before.)

So there you are. Nothing mysterious about it; just a couple of old friends used in a new way. Remember this and you're home free.

CATEGORIZING THE TROUBLE

Most of the time we get nice definite symptoms, like colored bars up and down the picture, green faces, and blobs of color floating around.

These troubles can be divided into two major groups. One, where there is no normal color in the picture, but there are colors on the screen, in blobs or bars. Two, where there is color in the right places, but it's the *wrong* color. Then there's a subgroup where the whole screen is tinted.

To get rid of this second one fast, it's usually the picture tube or an incorrect setting of a screen control that's causing the trouble. The set that came on a bright green, then delivered a good picture after about five minutes had a slow-heating *pair* of guns, red and blue. The green gun wasn't gassy, as we suspected. A heater–cathode short in one gun will cause a similar symptom, though this is usually permanent. This type of trouble can often be cured with an isolation-type "brightener" set to isolate only.

Screen-circuit voltage problems produce the same symptom. In another set, the green screen control wouldn't put the raster out at all. However, since we were able to get the other two sets to match it, and make a perfect black & white and color picture, we left it for the next time when it *had* to go to the shop.

Most of the true off-color problems are due to defects in 3.58 MHz oscillator phasing, distortion in the color demodulators, and odd defects in parts around the demodulators. All of them will respond to a little serious reasoning and the right interpretation of the clues on the picture-tube screen.

Intermittent Colors

In the intermittent-color or wrong-color department, look out for funny things happening in the horizontal oscillator/afc output section. This can affect the shape and size of the keying pulses used in the color section, since these come from the flyback. The important thing here is the *phasing* of the keying pulse.

Even though the picture seems to be fairly stable horizontally, it's possible for the pulse to be far enough off for the keying pulse to be out of range, which in turn upsets the burst-amplifier stage. This results in an intermittent or weak burst. Then you get a "*colors jump in and out*" complaint. Before making any other tests, be sure that the horizontal hold control is set as near to the center of its range as possible.

The fine tuning can also cause trouble here, for the same reason; and *aft* can, too. These circuits are simply discriminators; and incorrect alignment or a leaky transistor can make the fine-tuning drift, just the opposite of the intended effect.

In cases where colors are correct, but have a tendency to jump in and out very suddenly, look to the color bandpass amplifiers. These stages handle all color signals, and they are the most likely source for this complaint.

No Color—Odd Color Cases

One of the "*no color in picture but color present*" cases was a Zenith 20X1C38 chassis. The symptoms were a bluish screen, with two vertical color bars, red and green, about 1.5 inches wide, at the far left edge of the screen. No normal colors in the picture at all. To make things simpler, this was an intermittent condition. The black & white picture was slightly brownish, but with good detail, indicating that the Y channel was probably okay.

Turning the color control full on made the picture bright blue, with the vertical color bars much stronger. This also caused a very bad defocusing. Brightness and contrast controls worked. Tubes in the color circuits were changed, without results.

The dc voltages were odd. None were missing completely, as I'd hoped, but all were off. Some were high, some low. Voltages on the phase detector and killer diodes were well out of balance, which gave me a starting point. The 3.58 MHz oscillator was running. Color signal patterns on the R−Y demodulator were low, but the B−Y signal looked almost normal.

Fine-tuning reaction was normal, with color fringes (worms) showing. So the bandpass amplifiers (color amplifiers, in this set) were working. This confirmed the suspicion that the trouble was somewhere in or around the 3.58 MHz oscillator. The drastic imbalance in the control voltages could mean that the oscillator was running but was being thrown so far off frequency that it couldn't make normal colors, if indeed any at all.

Killing the burst and checking the reactions of the color oscillator showed that it was working. It had to be something in

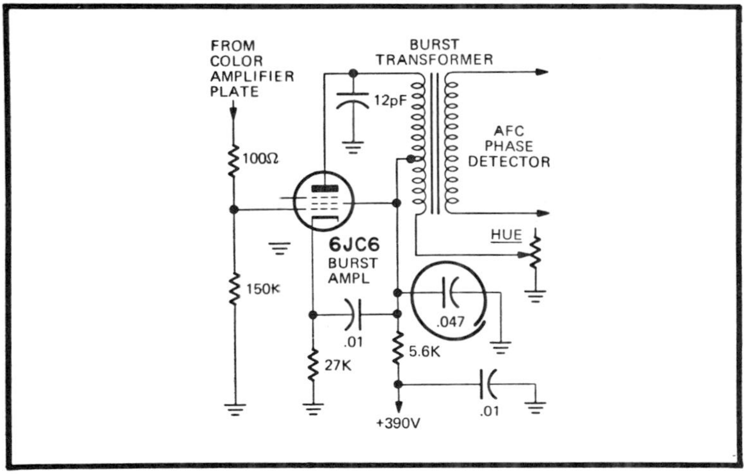

Fig. 1-2. Intermittent screen-grid bypass capacitor in burst amplifier of Zenith 20X1C38 caused all kinds of symptoms.

the burst amplifier. A little judicious punching and hammering around finally disclosed an intermittently open screen-bypass capacitor on the burst amplifier tube (Fig. 1-2).

This was apparently allowing the stage to develop some kind of parasitic oscillation that threw the burst off frequency and caused some kind of ringing reaction, with transient bursts of oscillation which made the colored bars on the left side of the screen, and the blue-screen symptom.

On a Zenith 20Y1C37 chassis, colors were present, but odd. Flesh tones just didn't "look right," and there was a certain amount of drift in the color. To make a long story short, this was finally traced to an imbalance in one of the 6ME8 high-level demodulator tubes. The dc voltages on the deflection plates of this tube were unequal (see Fig. 1-3). These should always be equal or within about 5V of each other. Plate voltages weren't too close, either. For a quick check on this, just swap the tubes. If the imbalance moves with the tube, throw the tube out! Check against the other tube. Incidentally, if both tubes are fairly old, it's a good idea to replace both of them. I changed only the unbalanced one, and had to go back a week later to replace the other one. (Same symptoms, different colors.)

Diode Grid Leakage

Some very odd color problems can be caused by a bad tube or crystal diode, if it's in the automatic color control

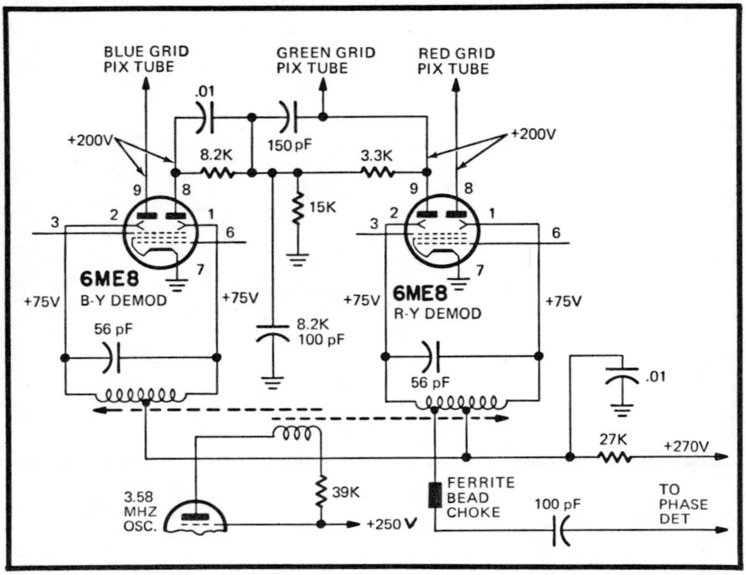

Fig. 1-3. Dc voltages on deflection plates of 6ME8 color demodulators should be equal.

stage. The diodes must be perfectly balanced if they're going to work. A lot of sets use the 6JU8 quadruple-diode tube. If this tube develops leakage between sections, look out! This is undoubtedly something like mount-contamination inside the tube, but it reads as *grid emission* on that type of tube tester. If everything else seems to be in fairly good shape, try a new 6JU8 before you start digging into other circuits, and most especially before you try any *realignment*!

Sparkle Plenty

In a GE KC chassis, colored sparkling and flashes were seen on the screen, mostly at the right and left sides. This eventually turned out to be a bad horizontal output tube, even though the raster was full width and there was plenty of high voltage. Seemed to be something similar to a Barkhausen oscillation, though there were no vertical bars on the screen. Try a new tube first.

Hybrid Video Amplifier Stages

Y-channel or video problems can cuase "color troubles." (In quotes because these aren't really color trouble, but they sure look like it.) This is especially true in sets with tube/transistor video amplifiers. We can have troubles that we don't expect if we forget that transistors can do things that tubes can't, such as shorting between input and output.

One of these is a shorted transistor that lets the video signal go through. However, it is slightly weak, and most important, it loses one phase inversion of the *signal polarity*. So we arrive at the picture tube with the brightness or Y-signal exactly opposite to what we'd like. In other words, a nice but *negative* picture.

You have never seen anything more peculiar looking than a good color picture with the video signal upside-down. Make it a habit to check for this; it's easy. Just turn the color control all the way off, and look to see if the black & white picture is negative.

Beside the shorted transistor, it is possible to develop shorts on the printed-circuit board which will have the same effect. Also, if the video transistor is a plug-in type, you can pull the transistor and accidentally reinsert it reversed. (I thought this would blow the transistor, but it didn't.) It will make a negative picture, though.

SMEARY COLOR

The black & white picture is clear, sharp, and with good contrast. Yet the customer is telling the truth when he says, "The color is a *mess*!"

Turn the color control completely off; good picture. When it's turned up to normal, "mess" is the only word for it. Large, smooth colored objects aren't too bad. Put a color-bar pattern on it, and this doesn't look too bad, either. Yet on a long shot of a crowd, where faces are about an inch or so in size, they show distinct vertical bands of red, green, and blue, followed by colors ringing, and colored ghosts all over the place. For another confusing symptom, the grid waveforms on the picture tube look almost normal.

Now you could be in deep trouble. However, if you know what to look for, where to look for it, and—above all—*how* to look for it, it's not too bad.

Step 1: Find out what *is* working by plain old observation: the tuner, i-f, video stages, picture tube, agc, brightness, color, tint control, *and* the 3.58 MHz oscillator. (The oscillator must be working or we wouldn't get any color.) This is definitely *not* a loss of color sync, either.

So what's left? The bandpass amplifiers, and perhaps the demodulators. The last one is not too likely, since the rocker pattern on the picture tube grids usually shows that these are all right. So we are left with a problem in the bandpass amplifiers. Now, *what* is it, and what *kind* of a fault is it?

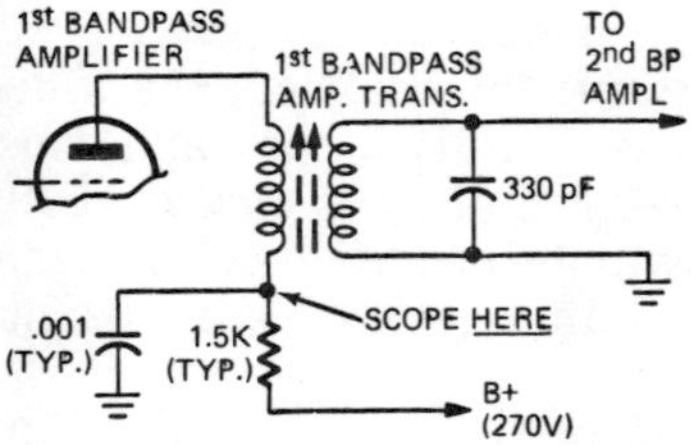

Fig. 1-4. Bandpass amplifier.

Reasoning will tell us this is not a short circuit. If it was, it would be more apt to kill the color signal entirely. So this is some kind of an *open circuit* (and not one in the signal path, either; we have signals). Something in the color circuits is *ringing*. This causes the miscoloration and the—most significant clue of all—the ghosts which are present *only* in the color signals. Turn the color off, and they go away.

I said *open* circuit because this must be caused by something which is open, and not affecting dc voltages. One very good suspect is a bypass capacitor, like the one at the bottom of the bandpass amplifier's transformer primary in Fig. 1-4. It puts the bottom of this winding at ac ground potential. If it's open, this sets up a feedback loop, the transformer rings, and away we go.

Simple test: Put the scope probe right on the bypass capacitor. At the high frequency of the color signal, it doesn't take a very large capacitor to bypass it. Somewhere between 0.001 and 0.01 μF is common. If you see *anything* at all in the way of a signal at this point, look out. The bypass capacitor isn't doing the job. Turn the scope gain up as high as you can. It doesn't take very much feedback signal to upset things for three feet in all directions.

Want to doublecheck? Bridge a good capacitor across this one, and see if it changes the symptoms. Should have no effect at all, if the junction is properly bypassed. Check all bypassed points like this. In some sets using transformers in the demodulators, check their bypassing, too. In one set, an open bypass on a demodulator transformer caused the screen to turn half gold and half blue.

Another problem is a defective winding on one of the bandpass-amplifier transformers. This won't be an open winding, in transformers which have only two windings without taps. This would break the signal path, and you'd probably have no color at all. This kind of problem will show up in circuits like those used in RCA CTC 39 and others, which

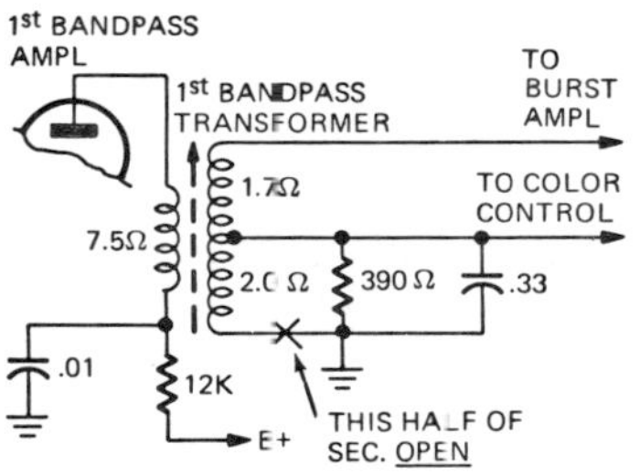

Fig. 1-5. Bandpass transformer secondary circuit.

have a tapped secondary winding on the bandpass transformers. One part of the winding can open, and the signal will still get through.

Since the normal impedance of the transformer is upset, it will ring. There's a very simple test for one of these; just grab the transformer with your fingertips. (Most of these are not shielded). If this causes the symptoms to change drastically, or even disappear completely, there you are. Check continuity of all windings, especially the secondary. Figure 1-5 shows one of these.

Also troublesome might be an *undamped* bandpass transformer winding. In some of the earlier sets, a damping resistor was connected across the secondary. If you happened to get the wrong bandpass transformer, it will fit the holes in the board, but the damping resistor will be hanging loose. Results; same ol' thing—transformer rings like a bell. Same check; grab the transformer between your fingers (see Fig. 1-6).

Although I haven't seen a verified case of this yet, there would be a definite possibility of trouble if the centertap of an output bandpass transformer secondary should either open or become ungrounded at a lug. The ohmmeter will catch any of these, as far as the open circuit is concerned.

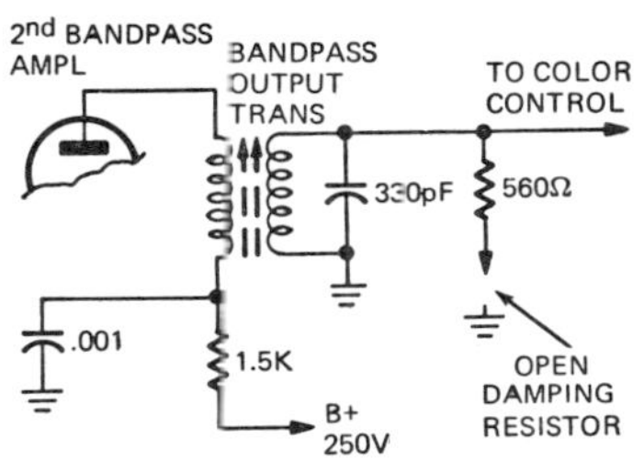

Fig. 1-6. Open resistor in bandpass transformer secondary circuit.

There is also a possibility that one of the electrolytic capacitors in the dc voltage supply could allow the feedback loop to set up through the power supply. Check these with the scope, or by bridging, or both.

These are weird troubles; yet, if you know what to look for, and where to look, they're not at all hard to dig out and cure.

CONVERGENCE CONTROLS

We've been seeing and using controls on convergence boards for a good while now, and there seems to be just a wee bit of confusion as to exactly *what* some of the controls are and what they do. The two most often confused seem to be *amplitude* and *tilt* controls.

We use parabolic waveforms in the convergence yoke, to get the results we must have; that is, good convergence all the way across (and up and down, too, of course) the screen. These change the convergence point of the three beams while they're moving. (After all, that's what "dynamic convergence" means—convergence while in motion.) To get exact compensation, we can't use a parabola that is exactly symmetrical. We might need more convergence at one place than at another. So we use controls on the waveforms.

One regulates the amplitude, just like a volume control, so this is the *amplitude* control. The other regulates the phase of the thing so that we can make one end or the other lift or drop; in other words, we can tilt the *waveform*. Now what will this do on the screen? Let's see.

We're trying, say, to get a red line to scoot in behind the blue and green that are already nicely overlapped. We see something that looks like Fig. 1-7. (The broken line represents the color that's showing too much.)

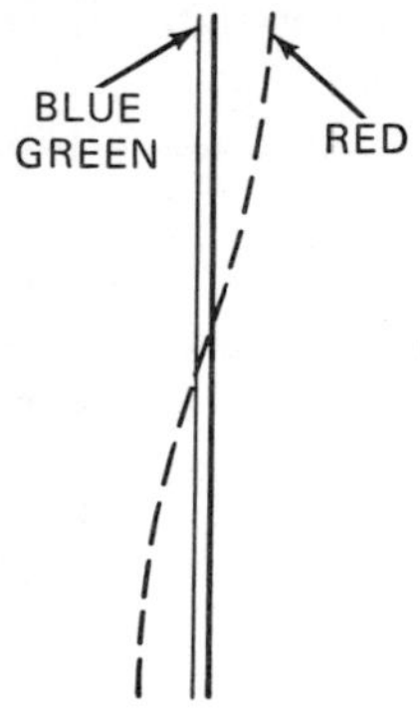

Fig. 1-7. Sometimes a color gets recalcitrant and just won't stay hidden.

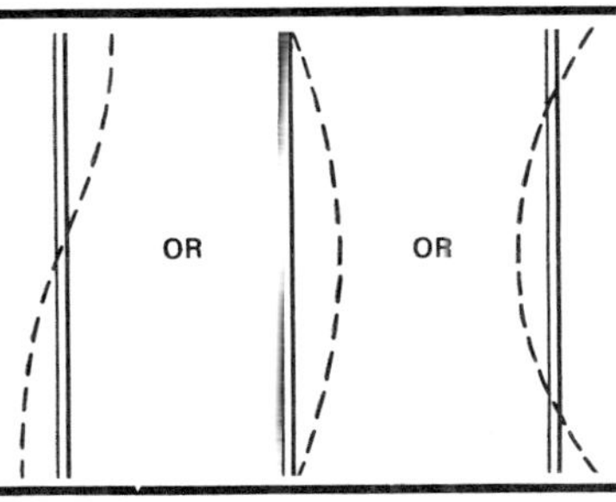

Fig. 1-8. Unparallel colors can't be straightened with amplitude.

Two of them are nice and straight. The other hangs out on one side at the top and the other side at the bottom! Now move the *red tilt* control, and you'll see the line go further out of convergence, going one way, and get closer to "right," going the other. Note that it stays almost stationary in the center! This is the *tilt* control. If we adjust it properly, we'll see this line straighten out and hide behind the others. Also we may see an opposite reaction: this line may bow in the middle, touching the others at the top and bottom! This is also a tilt-control reaction. For correct adjustment, we set the control so that the one we're adjusting is as near to *parallel* with the rest as we can get it.

Now what about the amplitude? Well, we may find that we have to deliberately introduce a bowing in this line to make it overlap the rest properly. So we set up our bow with the tilt control; perhaps it will be of just the right *shape*, but not cover the other lines, as in Fig. 1-8. To make it cover them, we add more amplitude by turning up the amplitude control. Once we get the line set up so that it is obviously the right *shape* (parallel to the others), we have the tilt set, and all we need to do is adjust the *amplitude* until it moves in behind the rest: And we get the white line we've been looking for all this time (Fig. 1-9).

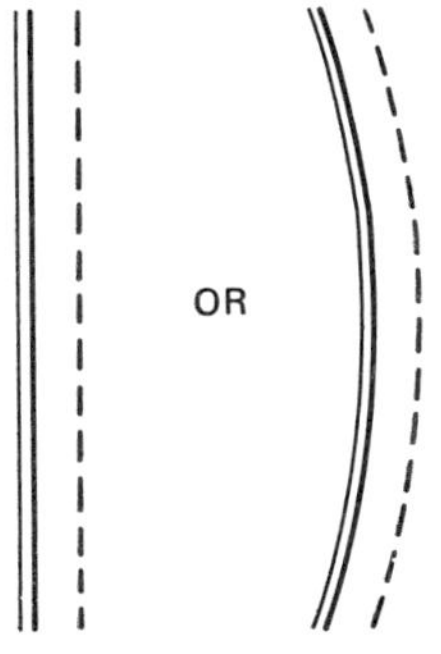

Fig. 1-9. The tilt control adjusts the shape; the amplitude adjusts width or intensity.

These are your clues: If a line is not of the right *shape*, adjust the *tilt* control; if it's the right shape but there's not enough of it, set the *amplitude* control.

COMB PATTERNS FOR COLOR TROUBLES

When we're looking for troubles in color TV, we need all the help we can get. This is especially true in hybrids and solid-state sets. We can't disconnect, check, and replace as easily as we once did. So we need a method that will pinpoint the trouble area fast!

Where we need most help is in the ones with "raster-but-no-picture" symptoms. Since the signal path in every radio and TV set ever built is a **series circuit**, all we need to know is "Where is it broken?" In other words, where does the signal stop or lose a lot of amplitude?

So back we go to one of the oldest test methods—signal-tracing. Of course, there's nothing like the scope for this. It shows you the signal. It also shows you the amplitude, distortion, and so on. For the quickest results we need a signal we can identify instantly.

We've got one. The color-bar pattern from the bar/dot generator makes an unmistakable "comb" pattern on the scope (see Fig. 1-10). The beauty of this is its simplicity. Just hook the bar/dot generator to the antenna terminals, turn the set on, and away you go. Follow the signal through the path until it stops. Then we go to the dc voltmeter and find out what's happened.

This can be used in any stage. For chasing signals through the video i-f stages, use a crystal demodulator probe. The waveform of Fig. 1-10 was taken at the collector of the 2nd i-f transistor in a hybrid Magnavox 936 color TV chassis. This one happens to be about 2.0V p-p. You do not have to read peak-to-peak voltage on every test point. As long as you can see an increase in pattern amplitude between input and output of a stage, fine.

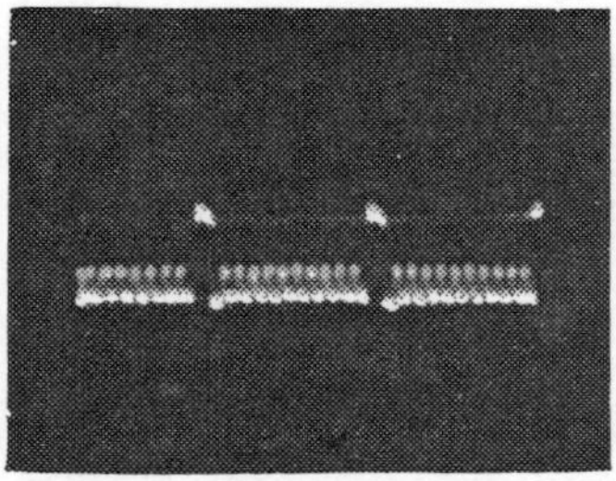

Fig. 1-10. Comb pattern taken with demodulator probe on collector of 2nd video i-f amplifier.

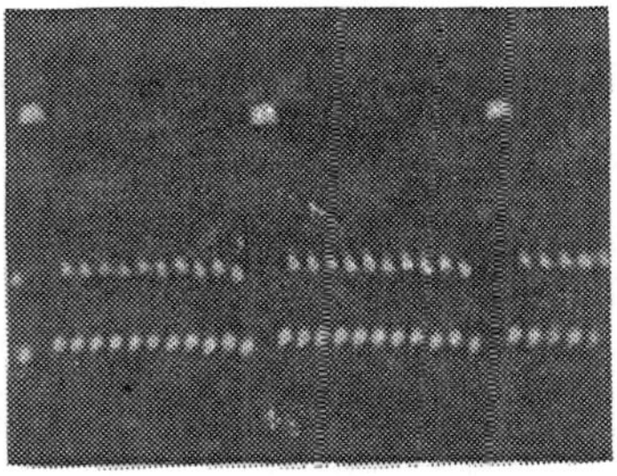

Fig. 1-11. Comb signal taken with direct probe at the base of the video driver: 1.2V p-p, sync positive-going.

If you want to take a shortcut, eliminating at least four tests, take your first reading on the video detector output with the *direct* probe. This should be at least 2.0 to 3.0V p-p and ought to look exactly like Fig. 1-10. If this signal is present at this point, you've cleared the tuner, video i-f, agc, dc voltage supply, and a raft of other things!

After the video detector, the low-capacitance or direct probes can be used. The same comb signal can be followed through the video amplifier stages. Look for changes in amplitude to check the gain. Starting with 2 to 3V p-p at the video detector, we should wind up with 250V p-p at the output of the video amplifier. If you see much less than that, you'll have a weak picture. Figure 1-11 shows the comb pattern taken at the 2nd video amplifier of the Magnavox 936 with a direct probe.

Obscure problems can be traced rapidly with this method. For example, there should be no signal inversion through an emitter-follower stage, but there should be as you go through a common-emitter stage. If you see the same signal, amplitude, and polarity at the input and output of a common-emitter stage, you've found the trouble. You'll also see a *negative* picture on the screen! If the colors look very odd turn the color off and see if the video isn't negative.

You do not need a wideband scope to make these tests. If the bandwidth and sensitivity of your old scope is a little limited you won't see sharp clean patterns, but you'll see'em. The waveforms of Figs. 1-10, 1-11, and 1-12 were made on a

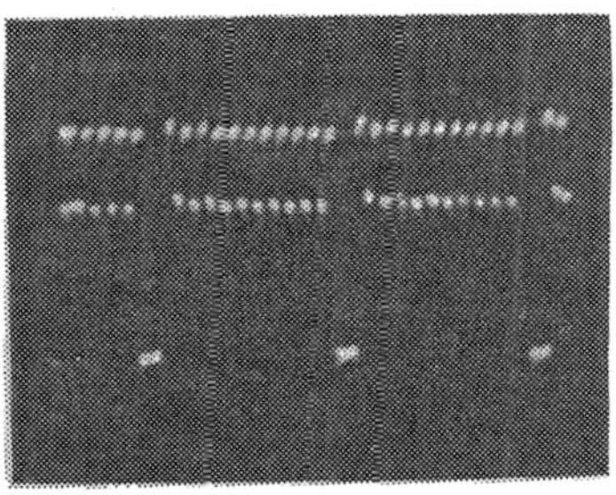

Fig.1-12. Same signal on grid of video output amplifier tube. 8.0V p-p inverted; sync negative-going.

scope old enough to vote. With the newer models you will be able to see signals of much lower amplitude; for example, the signal at the tuner output, which is only a few hundred microvolts.

With this method you can chase down trouble anywhere in the set—sync, agc, white-compression, sync clipping, and many others. The comb pattern has both vertical and horizontal sync pulses.

Using the demodulator probe, you can trace the color signals through the bandpass amplifiers, clear up to the demodulator input. Use the demodulator probe, for the color signals at this point are really rf; between 3.08 and 4.08 MHz—and AM at that. However, they will demodulate to show you the same comb pattern you see in other stages. Signal level at the input will be only a few volts, 1 or 2V typical. Output at the demodulator grid (input) will average about 30V p-p.

If you can see the color signals at this point, but still have no color on the screen, you can eliminate the bandpass amplifiers, the color killer (which generally cuts off the 2nd bandpass amplifier), and several other things. Go and scratch around in the 3.58 MHz oscillator or burst-amp circuits. The chances are that they're not working properly. A great many other "obscure" troubles can be found and fixed in this way, by eliminating certain things as possible causes.

AFPC TROUBLES

Color sync trouble or even a loss of color can often be traced to obscure faults in the afpc (automatic frequency and phase control) of the 3.58 MHz oscillator. Not really so obscure, of course, since there's no such thing as a fault which will not show you definite symptoms if you make the right tests, in the right places, with the right equipment!

Let's look at this circuit and see where these key test points are, what the normal reactions are. A typical afpc circuit is shown in the diagram of Fig. 1-13. The burst is fed into a centertapped transformer secondary winding. The signal is now 180° out of phase at each end of this winding. A diode is connected to each end, through small capacitors. The other ends of the diodes are tied together and matched resistors are connected across the diodes. The 3.58 MHz oscillator signal is fed to the junction of the diodes.

This is a simple phase detector. If it's working, you get a balance between the voltages developed across the diodes and a zero dc voltage at the centertap of the resistors when the

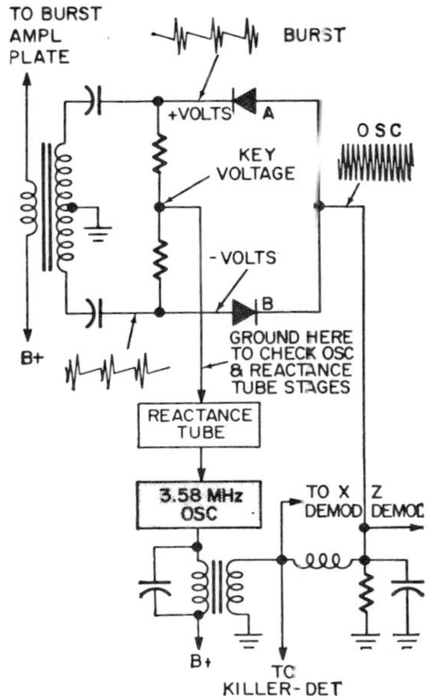

Fig. 1-13. Typical automatic frequency and phase control circuit, with waveforms.

color (or picture) is locked in. If it isn't working, the dc voltage is off zero and the control circuit pulls the oscillator *off frequency* instead of locking it in! Notice the direct similarity between color out-of-sync and a picture out of horizontal sync. Diagonal lines just as before, but now the *picture* stays on and the *color* breaks up into diagonal lines or rainbows of red, green, and blue.

Circuit Requirement

What do we have to have to make this work? First, burst. In this circuit, the burst comes from the burst amplifier plate through an isolated primary winding. In others, it may come to one end of the centertapped winding, with the other end grounded.

We must have a good stout burst signal in this winding, 180° out of phase at the ends, and that's all. The *amplitude* of this signal is important. Check the schematic to see what is normal for the one you're working on. Average values are at least 50V p-p, and sometimes more—up to perhaps 100V p-p.

Second, we need the 3.58 MHz oscillator signal. This is taken off at the phase-shifted Z demodulator tap of the oscillator output coil in this one. It may come from other points in other circuits, but it's got to be there. The presence or absence of either of these signals is easy to detect with a scope.

Right now all we need to know is "Are they there?"! Never mind the phase for a while. There's an awfully simple way of telling whether it's in phase or not. Just look at the picture!

Quick Checks

The fastest way, and one of the easiest tests you can make, is to read the dc voltage developed across each diode with a color-bar signal on the set's antenna input. You can do this on a TV signal, but the bar/dot generator gives you a steady, controllable input signal. This must be read with a high-impedance voltmeter. As you can see, the voltage will be positive on the cathode of diode A, and negative on the anode of diode B. These voltages *must be equal and opposite*!

If they're not, the circuit won't balance. Your voltage at the key test point, the centertap of the two resistors, will not be zero and the 3.58 MHz oscillator will be pulled off frequency. In normal operation, this voltage is zero. If the oscillator drifts one way, the phase detector develops a negative voltage and pulls it back. If it drifts in the other direction, the detector develops a positive voltage and pulls it back.

Weak Color Sync

If you have weak color sync, check these voltages. You may find that your reactance-tube grid is reading about 2.0V positive or negative. If so, this is wrong! The chances are that someone has run a setup adjustment on the oscillator, and pulled the oscillator back to a point where it will make color, by main strength and awkwardness. This is something like driving a car with one front tire flat. You can hold it on the road, but you have to exert a steady pull on the wheel; and if you let go for a moment, *pow*! The set will have a tendency to lose color sync on scene changes, station changes, etc.

Isolation Tests

Since you have a servo loop control system, you must break the loop and check one part at a time to check it reliably. Ground the grid of the reactance tube. This clamps it at zero voltage, or at the point where it ought to run. It also eliminates any afpc action; the oscillator is "free-wheeling."

With a color-bar signal on the screen, tune the reactance coil in the reactance-tube plate circuit. This varies the frequency (and phase) of the oscillator. If you can get the normal reaction on this test, the oscillator *is* capable of running on frequency. The normal reaction, as the reactance coil is tuned through its full range, is small, with numerous rainbows to start with, getting fewer and larger as you approach the correct point, then gradually straightening up and momentarily locking on the right colors (third bar red, in the bar pattern).

Going on in the same direction, it should fall out of sync again, making more and more rainbows. Some reactance coils will tune far enough to kill the 3.58 MHz oscillator or throw it so far out of phase that the color will drop out entirely.

Does this sound something like adjusting a horizontal oscillator with afc? It should! It's *exactly* the same method, substituting color rainbows for slanting lines in the picture!

Diode Tests

If the oscillator will set up and lock in momentarily, it's okay. If taking the ground off the reactance-tube grid makes the color fall out instead of locking tightly in, you've got diode problems. A voltage reading on the grid will probably show you a voltage quite a bit off zero.

Now, read the dc voltages on the diodes. These should be equal and opposite. This is easier to see on the grid test point, since you can use a lower voltage scale, preferably a zero-center. If you do not have a zero-center scale on your meter, you can set the needle to center-scale or somewhere near center with the ZERO adjust. Then you can see quite accurately whether the key voltage goes positive or negative.)

Normal Reaction

This key voltage should be zero. You should also be able to make it go positive or negative by adjusting the core of the reactance coil. The deviation should be the same on each side. For instance, ±6–8V seems to be a good average range.

If you find this voltage will go from about −8 volts up to −1 volt, then turns around and goes back toward negative again, you almost certainly have unbalanced diodes. The fastest way to check this is to substitute a matched pair of replacement diodes. Use the factory replacements if possible. If both voltages now read the same *polarity*, turn it off and reverse one of the diodes!

If New Diodes Won't Help

If the replacement diodes won't make the voltage zero, try a new crystal. This is not common, but in a few cases the resonant frequency of the crystal will be off just enough to make it impossible to balance the circuit properly. If the oscillator frequency seems steady enough after it is adjusted, but the voltage is not correct and won't "cross zero," this would be the next thing.

All of the little capacitors in this circuit should be checked, of course. One that causes some odd problems is the feedback or neutralizing capacitor from the plate to grid of the reactance tube. In the set from which this was taken, this is a 3.9 pF NPO. Any type can be used for testing, but for replacement, an exact duplicate must be used. The temperature coefficient is very important; this one is an NPO, meaning zero drift in capacitance with temperature change.

The Killer Detector

Another circuit, not shown here, is fed from the same transformer winding. This is the killer detector. It uses the same diodes, connected in the same way, with the pair of resistors, etc. However, this one is fed 3.58 MHz signal from a different point, at the X demodulator takeoff. This is 90° (in this one) out of phase with the Z demodulator takeoff.

This circuit is *intentionally* unbalanced, and it will develop a high negative voltage at the junction of the two resistors. It is used to signal the *presence of burst* in the incoming signal, and makes the color killer open the color channel and let the signals through.

This is not as critical as the phase detector, of course, but it can cause trouble. If you have "no color" problems, such as the killer cutting off the bandpass amplifier on color programs, check these diodes. If one of them should open, for example, it could upset the killer bias.

If you know what this thing does and how it does it, it's not at all difficult to fix. Just pretend you're working on a plain old horizontal afc circuit! It'll be a lot easier.

READER QUESTIONS

Poor Color Lock

I've worked over this Magnavox T920 color set pretty well and it runs fairly well now. Only problem is a "loose" color sync. If I flip the tint control back and forth, I can get it to lock and it stays in for quite a while. What could be the cause?—H.W., San Jose, CA.

There are a couple of things. One is a slight imbalance in the afc diodes. Check for equal dc voltages, opposite polarity. Should be somewhere around 50V.

Another and sometimes more puzzling thing is a crystal that's just a wee bit "off." Good check for this: read the dc voltage on the grid of the reactance tube with a color-bar pattern on the set, and locked in.

The voltage here should be *zero*. Now move the core of the reactance coil about one turn in either direction. You should be able to make this voltage *swing* positive one way and negative the other. With a perfect crystal, the dc voltage here will swing about 4 to 5 volts in each direction. If you find that the voltage will swing about 8 to 9 volts in one direction, but will not "cross zero" and reverse the polarity, change the crystal.

Sudden Loss of Convergence

Here's one for you. In a Sylvania D-15, I had a sudden loss of convergence. I couldn't get the center vertical line, red and green, to converge at all. The trouble turned out to be a shorted electrolytic capacitor, C-800 (6.8 μF), located on the convergence board. This might apply to other chassis, too. Harold's TV, 810 College, Bowling Green, KY.

Thank you very much, Harold. Filed and noted for future reference.

Color Sync Lost

This Motorola TS-918 had a bad arc-over from the ultor connection to the shield; burned up the insulating cap. After I got this fixed, I found there was very little color sync, and the color looked very odd. Should I realign the color stages?—R.G., Provo, UT.

No. The chances are that this is component damage, due to the high transients generated by the arcing. I have seen similar things happen.

In this model, check the color sync and gate stage, the 6BL8. (See Fig. 1-14.) It is quite possible that the 4700-ohm glass-film resistor has been damaged. If so, you'll have no gating pulse on the 3.58 MHz signal to the crystal and no color sync at all. This is R931.

Colors Reversed

This Zenith 16Z7C17 shows a complete reversal of the colors. Blues on the left, green in center and red on the right. Color bars also miss the black-and-white bars by about half, falling in the slots. I've checked and aligned the color circuits, and everything I can think of, with no results. What is it?—L.G., Chicago, IL.

A phasing problem. Either the color signals or the reference oscillator must be way out of phase. This is the only thing that could cause such a tint shift. Check, by replacement, that little phase-shift coil in the 3.58 MHz oscillator output.

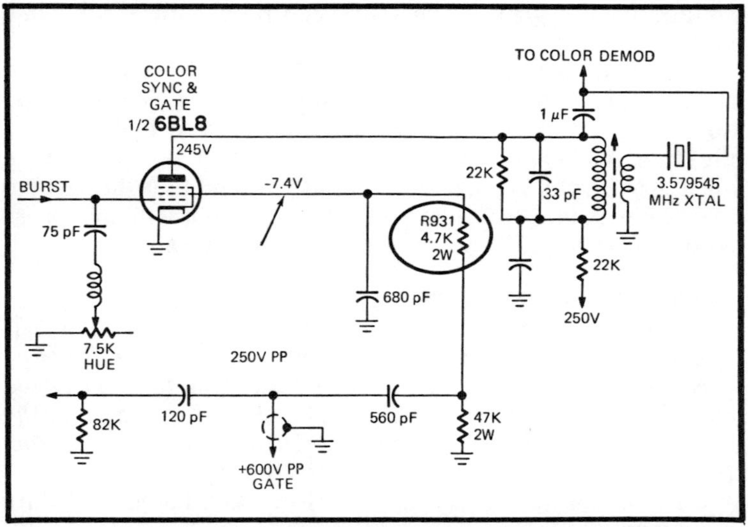

Fig. 1-14. Color sync and gate circuitry, Motorola T5-918.

If it is shorted, it can upset the normal phase shift very badly. It has been known to do this. Try a new one.

Odd Color Problems

This 20X1C38 Zenith has intermittent "odd-color" problems. I notice on a tube-test that the 6JU8 tube has quite an imbalance between sections.—T.D., Franklin, MI.

You've got it. Replace that 6JU8. Imbalance and other odd troubles, including mount contamination, in this tube can cause more weird troubles than you can shake a stick at. This should fix it. (Field feedback from a reader: it did.)

The color keeps popping in and out and acting very funny on this GE CB21 chassis. All tubes in the color section have been replaced; no help. Where do I go from here?—H.S., Vincennes, IN.

After all of the tests you have made (which were right!) I believe I'd go and have a look at that little neon lamp in *series* with the burst amplifier grid. This has caused some intermittent color problems in the past. Try a new one.

Band of Ripples

I've got some of the most weird symptoms I've ever seen in this Motorola Quasar II TS-929 chassis. With a picture, there's a vertical "band of ripples" right in the middle of the screen, flickering with all colors! If I take the antenna off, I get very thick black vertical bars, odd shaped, like

Barkhausen, but definitely not; they're all over the screen! I scoped the horizontal sync, and the sync-pulse goes all to pieces when the antenna is disconnected.—R.T., Hatfield, AR.

I've seen this before. Uncommon, but it happens. Look for a very small rf choke, encapsulated, mounted on the plate cap of the 6LF6 horizontal output tube. Take a piece of wire and short this choke! This will probably clear it up.

Apparently, this choke causes the horizontal output stage to ring heavily at some high frequency, and generate parasitic oscillations. Shunt it, and then quit!

Intermittent Convergence

We've got an odd case of intermittent convergence in a Magnavox 938. The customer complains bitterly, but we had an awful time making it show up in the shop. When we did, it looked like a sudden displacement of the three colors, by about ½ inch each —T.D., Bellevue, OH.

Check the Molex plug on the convergence yoke for intermittent contacts or loose wires to the male pins. Check the 50 μF electrolytic in the vertical output tube cathode circuit. If it is opening intermittently, it can cause some strange symptoms. While you're at it, hit the diode unit on the convergence board a couple of smart raps with a screwdriver. These can "loosen up" and cause just this type of trouble. If it's loose, replace it.

No Red Setup Line?

I have two "funnies." On a CTC 19 RCA chassis, I get no red line on the screen in service position. Yet, I can get a bright red raster for purity, black & white picture good, color pix good. Pix guns check out okay on tester.

I've also got an intermittent loss of the video signal at the 12HG7 plate, but plenty at the grid. What gives?—J.B., New York, NY.

For the intermittent loss of video, check all of the solder joints on the board between the 12HG7 plate and the pix tube cathodes. Peaking chokes, etc.

For the *no red line* symptom: Forget it! Since you have a good monochrome picture, all three guns can conduct properly. If the color picture is good, this clears all color circuitry.

Color Pops In and Out

The color is intermittent on this Zenith 12B8C15. Acts like color killer trouble, but this one doesn't have a killer. I can rap on the chassis near the IC demodulator, and make it act up.—A.M., Philadelphia, PA.

Check the IC socket; if it is tight and clean, I believe I'd try a new demodulator chip. This has been the cause of this kind of trouble in some of these chassis.

Blanked Color

This GE H-3 portable color TV has a peculiar problem. According to the owner, serveral months ago the color started disappearing on the right side of the raster. Now, it gets color only on the left ¼ of the screen! The monochrome picture is good over the whole screen. Color and tint controls have what seems to be a normal effect on what color there is. I'm puzzled!—M.M., Los Angeles, CA.

So am I. Let's rub the crystal ball and see what we can see. Obviously, your *blanking* circuitry is overdoing things just a little. In this chassis, blanking is fed into the cathodes of the 6AC10 color difference amplifiers. So, you're actually blanking only the color.

It sounds as if the blanking pulse is far too wide, or badly distorted. Check the coupling capacitor between the blanker and the 6AC10 cathodes, as well as the blanker tube and 6AC10 itself, and all resistors.

Color Blooming

This Zenith 20X1C38 has an odd symptom. There is little or no blooming on a monochrome picture. However, on color programs, it blooms. The higher the color control is set, the worse the bloom. Everything seems to be normal, except that the picture tube screens have to be set about 300V higher to get normal brightness, when the service switch is returned to normal. What causes this?—R.H., Aurora, IL.

I think you just told me. I started to say "Check to see if the picture-tube screen controls aren't set too high." This will often cause color blooming or plain blooming. This would be necessary, if you have some other fault which makes it necessary to run the screens up to get a picture.

The most likely place for this would be in the picture tube grid circuits. If these voltages are too low (too far toward negative) this would reduce the beam current; this could be "corrected" by advancing the screen controls. However, the real fault will be in something that affects the grid voltages on the picture tube.

Since all colors are affected, check the *supply* voltage for the grid circuits, including the tubes. You'll probably find a dropping resistor that has increased in value, or something like that.

No Color, No Picture

The picture on a Zenith color set looks very funny. When I turn the color control off, I get no picture at all! I'm just starting in color, and I've never seen this one before!—D.A., Babbitt, MN.

The "picture" in a color set is made up of *two* parts; the color and the video. Color signals are applied to the grids of the crt, video signals to the cathodes. In this case you have obviously lost the video signal completely!

Check the video output amplifier. More likely cause, check the video driver transistor. Incidentally, if you find one of these with a *negative* black & white picture and some really odd colors, this transistor is probably in its socket *backward*!

Autodegausser and Purity

I've a Zenith 20Y1C38 with several problems. The complaint was bad purity, and fuse-blowing at turn-on. I replaced the thermistor (cracked) and the degaussing coil, which has a burnt spot. Helped fuse-blowing, but purity still bad. Still blows a fuse now and then. What's going on?—E.E., Baltimore, MD.

I think it's not what's going on, but what has gone before. This set has been visited by lightning! (Very common cause of this kind of damage.) This is what cracked the thermistor,

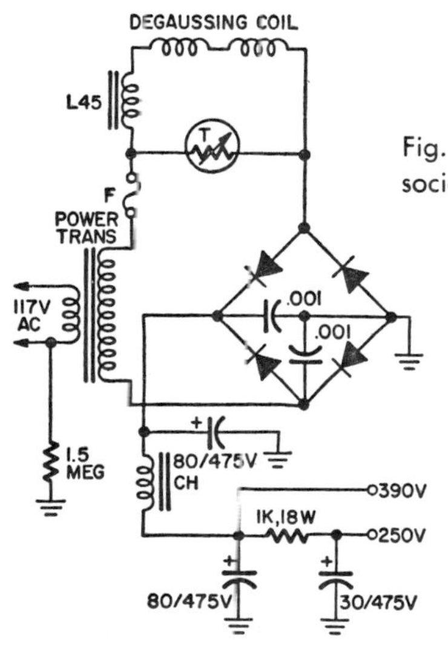

Fig. 1-15. Autodegausser and associated circuit, Zenith 20Y1C.

blew the hole in the degaussing coil, etc. There may have been other damage too.

I would recommend checking or replacing that small choke in the degausser circuit (L45, Fig. 1-15). It could have shorted turns or an intermittent leakage to the core, which would affect the degaussing current; hence purity.

Try this: Disconnect the degaussing coil, turn the set on, and degauss the whole tube very thoroughly with a standard degaussing coil. Check the purity. Turn the set on and off several times, cooling it off about 5 minutes between tests. If the purity holds, reconnect the autodegaussing coil and repeat these tests. If the purity *goes off* with the autodegaussing circuit connected, replace that little choke. It's the only part left!

One more possibility; rare, but has been found to happen. Check *all* the rectifier diodes in the bridge. If one of these was blown *open*, your output voltage won't be affected too much, but your *ripple* will have a very *peculiar waveform*! This can upset the action of the degaussing coil, and cause it to "regauss" the tube instead of demagnetizing it.

Color Alignment Query

I have a question on color TV alignment. A friend insists that besides taking out the horizontal output tube, you must add a load resistor from B+ to ground. I note that most companies do not show this any more. Is it absolutely necessary?—A.M., Morgan Hill, CA.

The purpose of this load resistor on B+ was to hold the voltage down to "normal." RCA used to insist on this, but they don't any more.

The voltage regulation of the B+ supplies in practically all color sets should be good enough to hold the voltage well within limits. You can check for this with a dc voltmeter. If the voltage doesn't jump more than 25–30V, all is well.

Convegence Troubles

I've got a convergence problem that won't quit! All waveforms on the convergence board look fine, as nearly as I can tell. But the lines bow on me, and I can't straighten them out. What's going on?—G.S., Peoria, IL.

If your waveforms look okay at the input to the convergence board and you still have a very severe bowing of the lines (especially horizontal lines), check the clamping diode assembly on the convergence board. If one of these diodes is leaky or shorted, this will cause a very severe bowing of the line of that color.

The purpose of the these diodes is to "clamp" the dynamic convergence sweep *at the center* of the screen, giving it more effect on the beams at each end of sweep. So if the diode is bad, the entire line will bow, and won't respond to the controls. Quick check: just short each diode with a screwdriver. This should *cause* bowing. No effect, this is shorted!

Horizontal Hold and Color Burst

Why does the horizontal hold control affect the color on my set, and on quite a few of those I work on? The picture will hold, but the color will fall completely out of sync.—J.W., Portland, OR.

This puzzles a lot of people, but it's normal. The color burst is *gated* by a pulse from the horizontal circuits. Unless this gating pulse is *exactly* (within about 5–7 μ sec) in phase with the color burst, the gate won't open. (See Fig. 1-16.)

Red Faces

I've got an odd problem in a Sylvania D-12 color chassis. I get color, but faces are very red. The color control works, and makes them redder. The tint control shifts the bars on a color-bar pattern pretty well, but on a color signal it just makes faces redder, too. Black & white picture okay.—R. D., Las Vegas, NV.

Check your demodulator transistors. I think you'll find one of them is bad. Since you have a good black & white picture,

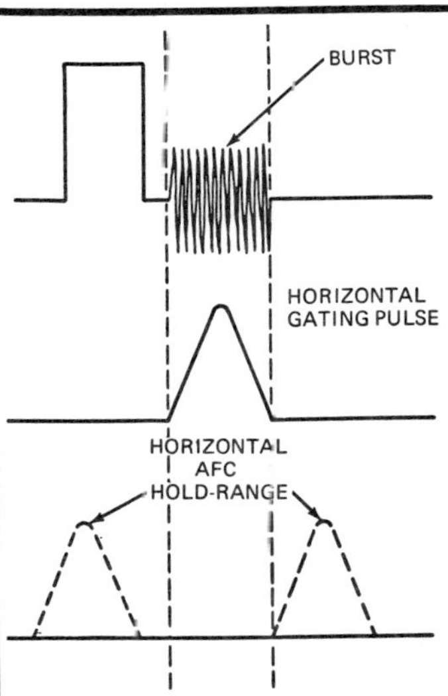

Fig. 1-16. The color burst (top) occurs at a specific point in the video waveform. The gate for this burst is a pulse (center) which must occur during the period burst is present. The afc will compensate for some noncoincidence of signals, but its range, of course, is limited.

this shows that your color diff-amps are working at least. (If one of these was bad, you'd have a tinted picture; reddish, bluish, etc.) However, incorrect demodulation can produce this effect. If one is bad, change them both, and check to make sure they're matched.

Discolored Picture

I thought this was a color problem, at first, but it doesn't seem to be. On this Zenith 14A9C50, there are 3-inch "circles" of different colors all over the screen! If I turn the color control off, they stay. I suspected the convergence yoke; unplugged it, and got the same kind of pattern.—H.F., NJ.

Try this: degauss the tube thoroughly, with an external degaussing coil. Now check to see if the colored circles have disappeared. I have a notion they will. If so, then turn the set on and off several times, letting it cool for a few minutes each time. If the discoloration comes back, you have a defective autodegausser in the set.

Hum Bars In the Color

CTC 11, RCA, shows horizontal bar about 3 inches high. This creeps up the screen slowly, then repeats. This symptom is present only on color. A solution would be appreciated.—F.M., Blauvelt, NY.

We can eliminate quite a few stages right away. The dc power supply (this is a 60 Hz ripple, not 120 Hz), the video, and color difference amplifiers (these work on black & white as well as color), and so on. This is caused by some stage which handles *only* color signals. It is most apt to be due to heater–cathode leakage in some tube which fits the specs and has a cathode bias resistor!

This would be: (1) the 6AU6 color bandpass amplifier; (2) the 6AL5 color phase detector; (3) the 6AW8 killer/1st video, and (4) the 6GH8 3.58 MHz oscillator and control. Try new tubes in these sockets, one at a time.

Intermittent Loss of Red

The reader wrote: "I am getting an intermittent loss of red on a Zenith 14A9C19 chassis, with the chip demodulator." I replied: "Check the signal-in vs the signal-out of the demodulator chip. If it is only on the one color, and the rest work, then I'd suspect a bad chip. Be sure that external parts are definitely good."

He replied: "It *was* a defective demodulator. For your information, the Zenith distributor says that the recommended replacement for this chip is the "silver" type, part No. 221-39, to replace the original "black" unit, part No. 221-37."

Slow Color Warmup

It takes about 20 minutes for the color to come on in this Zenith 14A9C50 chassis. I'm enclosing a list of the dc voltages around the color stages. Any ideas?—L.T., Winnipeg, Man.

One at least. You show −0.5V on the color amplifier grid (pin 7, 6KT8) which is normal with no signal or a black & white signal. However, when a color signal is received, this should *change* to a +6.0V. It doesn't. So, your color killer circuit isn't working or isn't starting up as it should. The color amplifier stage is being cut off, either at all times or for far too long.

Check the acc/killer diodes. The dc voltages on these should be equal and *opposite*, and should change when a color signal is received. This will run about −50V on the anode of CR213, and +50V on the cathode of CR214.

Also, check for the presence of the 3.58 MHz oscillator signal. In a few sets, this oscillator is slow starting; weak tube, bad crystal, etc. This will do the same thing.

Too Much Green

The screen of this RCA CTC 16 is green all the time; color and black & white. The dc voltage readings on the G−Y amplifier are okay. B−Y and R−Y plates read +80V. The pix tube is good. Resistors and capacitors in the color-difference amplifier circuits all check.—T.C., Cicero, IL.

You've found the problem: that low plate voltage on the color difference amplifiers. Since this determines the conduction of the pix-tube guns by controlling the grid voltages, the "less-positive" voltage here shows that these two grids are almost cut off (in other words, the picture-tube grids are "too far negative," which reduces the conduction of any tube).

Most likely cause for this; a bad solder joint on the printed-circuit board at the B+ supply point for the two 27,000-ohm plate resistors of the B−Y and R−Y amplifier stages. These are the 2W light blue resistors just behind the 6GU7s. The G−Y plate resistor goes to a different connection.

Loses Color Purity

I've got an odd problem in a Motorola TS-914 chassis. It loses purity as it runs. After it's been on about 2 or 3 hours, I get blobs of color on the screen. These are hard to get out even with an external degausser. The degauss circuit seems to be okay; I read 25 ohms on the coils. Where do I go from here?—C.C., Dix Hills, NY.

This sounds as if it is a problem in the set's autodegausser. This set has an unusual degausser circuit, operated from the focus voltage; when the *degauss* button is pushed, a charge is sent through the coil (from a capacitor).

If you read 25 ohms on that coil, look out. The factory schematic shows a normal resistance of only 1.1 ohms. This could indicate a bad degausser coil. Check the switch and capacitor for leakage. Incidentally, you cannot run this set if the degausser coil isn't hooked up; it upsets the focus.

Oversaturation

The picture comes on with the colors oversaturated, and well out of phase. Raster's clear; no impurity. This is a Zenith 14A9C50. I ran an afpc adjustment, and replaced the IC demodulator chip and the oscillator tube. No help. I'm lost.—G.C., Rockford, IL.

This sounds like a 3.58 MHz *crystal* which won't quite come to the right frequency! Hook a 50K resistor in series with the probe of a high-impedance voltmeter and read the dc voltage on the grid of the 3.58 MHz oscillator tube. Use a color-bar signal on the set, and set up for afpc alignment.

Normally, the grid voltage should be zero. It should be able to go from a −5V to a +5V as you adjust the oscillator frequency. If the crystal is off, it will probably come up to about +0.5V or −0.5V and then snap out. Final test is to try a new crystal.

Red Flicker

I've got an odd problem in an Admiral 25E6 chassis. Colors flicker at a fast rate; seems to be mostly in the red. Otherwise okay; this is not loss of color sync or barberpole. Haven't been able to pin it down.—J.H., Los Angeles, CA.

There's one way of finding out. Eliminate the other two colors! Turn the *blue* and *green* screens off, and check it on a pure red picture or pattern. This type of trouble is often due to things like an intermittent solder joint, peaking choke, capacitor, or resistor—in the R−Y color-difference amplifier stage, of course.

Another good possibility: Check the wiring of the pix tube *socket* for dirty contact, broken wires, etc. "Tap around" in the R−Y amplifier circuit with a pencil, looking for loose contacts.

Color Problems, One Channel

The color is okay on all channels but 13, in this GE G13 chassis. Sound and picture are off, too; if I tune for one, I lose the other. No color, except when I balance the channel selector between 12 and 13.—H.J., Fresno, CA.

From the symptoms, I'd say "tuner." Everything else is common. If you'll sweep the rf response of ch. 13 on this tuner, you'll probably find something that looks like a camel saddle! Single-channel problems with these characteristics are almost always due to tuner troubles, since this is the only thing that is "changed" when you change channels.

Blobs of Color, But No Color

This Magnavox T-924 chassis came in for service with a very small picture, which soon disappeared. Checking the power supply, I found the thermistor and varistor in the autodegausser completely gone; just 4 wires with blobs of metal on the end. After replacing these, and one of the power supply diodes, I got a picture.

However, I have no color. The screen is covered with big blobs of different colors, but they're stationary. Color and tint controls have no effect. What else is wrong here?—N.C., Brooklyn, NY.

This kid has been tapped by lightning, and hard! When this happens, the resulting tremendous transient will *magnetize* the picture tube, due to the degaussing coils themselves.

Degauss the tube very thoroughly with a big coil, and I think your color will come back. The actual problem here is that the tube is so badly "out of purity" that none of the colors will register.

No Color Sync

I've got no color sync in a Magnavox 933 chassis. When I ground the grid of the reactance tube, and adjust the reactance coil, I get a weird effect! I can get rainbows, then tune the coil slowly toward the point where the color ought to lock in, but instead, at that point, it drops out entirely! This does the same thing on either side of what ought to be the resonant point! What is this?—R.T., AR.

Check your dc voltages on the acc and killer diodes. I expect you'll find them *unbalanced*. You should get exactly the same voltage, with opposite polarities, on the two diodes in the acc. If they won't balance, replace them with a matched pair. Same thing for the killer diodes. Acc diodes should read about 13 volts; + on one and − on the other, but they must be able to balance, for the color oscillator works with exactly zero volts on the control tube grid; + going one way, − the other.

Floating Bars In Color

In an almost new solid-state RCA CTC 44B chassis, we have good color, but slowly rising "bars" (horizontal) of red, green, and blue, about 2 in. wide, through the picture. There's also a pretty bad hum at high volume, which is reduced when the treble control is turned full up. Once in a while, the channel-change function of the remote control acts up, too.—F.D., NY.

With so many different things going on at once, it must have a common cause! Look for something like an open filter capacitor. (This was the original answer, cut down; the reader came back at us later. He found an open 200 μF electrolytic capacitor, C1109, in the remote control receiver!)

Purity Problem

After changing the picture tube in a Motorola TS-914 chassis, I ran into an odd purity problem. We could adjust it for an all-red raster, turn it off, and when it was turned on again it wasn't red any more. Readjusting things didn't help! What gives?—H.S., Vincennes, IN.

Check the degaussing switch. It sounds very much as if the tube is being "regaussed" instead of degaussed. This model used a thermal switch; if the contacts arc and weld shut, the degaussing coil stays in the circuit and actually remagnetizes the tube.

Check; degauss it with a regular coil, then turn it on and off several times, letting it cool between times. If the purity gets steadily worse, check that switch. Replacement is the best cure.

Convergence Drift On Channel Change

I got a screwball symptom! When I change channels on this Zenith 23XC38, the convergence drifts off! After about 3–5 seconds, it slowly comes back to normal; shows up worse on a dot pattern. Even if you just move the channel-selector off-station then back, it pops out of convergence then slowly comes back.—R.C., Gorham, ME.

You know, I'm kinda sorry you asked! Seriously, though, this seems to be due to the "switching transient" caused when you change channels. From the symptoms, you have an electrolytic capacitor somewhere in there that is either charged or discharged, and takes 3–5 seconds to recover. During this time, it evidently affects the convergence waveforms.

I can think of one capacitor in this chassis that will do just that. It's the big electrolytic in the cathode of the vertical output stage. This looks like an ordinary cathode bypass, and is, but it has another function. It feeds dynamic convergence pulses into the convergence board.

So if this capacitor is intermittent, as many of them are, and can be "shocked in or out" by a transient, this could be the cause of the drift. I do know from personal experience that trouble in this capacitor—high power factor, or intermittent opening—can upset dynamic convergence. Try a new one in there. If that doesn't do it, look for other coupling capacitors which feed pulses to the convergence board.

Vertical Bars In Color, RCA CTC 28

In an RCA CTC 28, there are about seven vertical bars on the screen, about ¾ inch wide. At the left side of the screen, the red seems to be missing. On a weak signal, this is worse. I've checked out the antenna, all of the horizontal section tubes, etc., with no luck. Swapped the 6GU7 tubes; no help. Can't see anything that could cause a 60 Hz pickup.—K.W., Fullerton, CA.

This isn't a 60 Hz problem; that makes *horizontal* bars. The basis of this trouble is some sort of stray pickup of the original *horizontal blanking* waveform. In this, the spike does the blanking, and the ringing along the baseline must be completely clipped off! I believe that these ringing waves are the cause of this trouble.

Check the waveforms on the horizontal blanker, the "1–2–3" triode of the 6GU7 (other part is the G–Y amplifier). Also, check the keying pulse on the burst amplifier grid, which also goes to the 6KE8 color bandpass amplifier grid. If the ringing is too great on this waveform (see Fig. 1-17), it can override the clamp action of bypass capacitors, etc. and get into the color or burst amplifiers. There are several bypass capacitors which could open and allow this to happen. Try shunting each of these with another capacitor. Since they will all cause the same trouble, we can't pinpoint it without making actual tests.

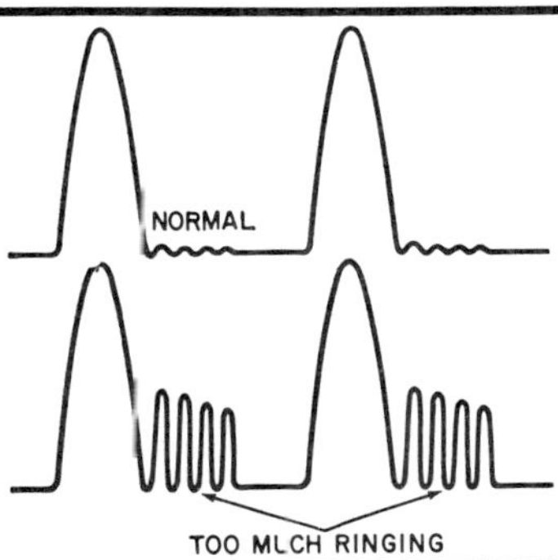

Fig. 1-17. The waveform on the horizontal blanker should have an almost flat baseline; if "sub-pulses" are evident, the excessive ringing could cause multiple symptoms, including vertical hum bars. An open bypass is a likely source of ringing problems.

Chapter 2 Sweep & Deflection Problems

One of the most mysterious troubles in horizontal output stages is the "creeper." The set plays perfectly when first turned on. Then the cathode current of the horizontal output tube starts creeping up, and up, and finally—click!—out goes the circuit breaker or the flyback, if the set doesn't have the proper protection. In most cases, the raster will pull in, dim, lose focus and gradually disappear.

What causes this? In most sets the standard tests will show no high leakage, or similar problems, in any of the numerous "loads" on this circuit. By loads, I mean the high voltage, boost, boost-boost, sweep, and focus. In some, all tubes will be new. So, what can cause such a symptom?

There are quite a few suspects. Normal suspects, that is. For one, a gradual loss of grid-drive signal. If the horizontal oscillator is weak, and the peak-to-peak voltage of the drive gradually falls off, you'd see the same symptom. With low drive, the grid voltage of the output tube goes more positive, and cathode current goes up. This voltage is developed by grid-leak action in the horizontal output tube's grid circuit.

A slight leakage in the coupling capacitor might be a good suspect. However, capacitor leakage is more or less fixed, and won't show the creep symptom. They're seldom thermal, too. But don't take chances; check it.

Another dandy cause for this is the grid-bias control type of high-voltage holddown circuit. These things use a pulse from the flyback, fed into a diode in the horizontal output tube's grid return circuit. When the output goes up, the pulse amplitude increases. This is rectified and converted into a higher negative bias for the output tube, holding down its output. There are usually several high-value resistors used in this circuit, always a good suspect for thermal drift. Diode leakage, too, could do it, although this is rare.

All of these are more or less "normal" causes, and we check them as a matter of course. However, a new one has been cropping up of late, and I thought it would be a very good idea to bring this out. It can be identified immediately, by a very simple test, *if* you know about it. (To be frank about it, I discovered it while looking for something else.)

The basic fault is *grid emission* in the horizontal output tube! When this happens, the tube starts drawing grid current. This makes the grid go more negative, due to the grid-current flow through the high-value grid resistor. So instead of conducting for the proper amount of time (which is very short, something like 8 to 10 μsec), the tube stays on longer. This increases the pulse width and conduction time and with it the average cathode current. This phenomenon is definitely thermal; the longer the tube operates, the greater the grid emission and the greater the heat dissipation.

The "simple" test? Just hang a scope probe on the control grid of the horizontal output tube, and watch the waveform. The normal waveform is like Fig. 2-1A, with a nice sharp peak. When this fault develops, you'll see this peak gradually start to flatten out, as in Fig. 2-1B. Even a flattening of this amount can raise the cathode current quite a bit. When it has flattened to something like Fig. 2-1C your cathode current will be up around 400 mA, and the breaker had better be getting ready to trip. The equivalent pulse widths are shown below, so you can see why it acts like this.

The cure? Simple: *another* new tube. Cook the set, with the scope still on the grid and the meter in the cathode. In one of the first cases I ran into, a brand new set showed these symptoms. The horizontal output tube was replaced, only to find exactly the same trouble. A third new tube turned out to be good. The first two were checked on a tube tester capable of reading grid emission, and showed a very high reading. There emission was good, of course; any tube that can draw 400 mA is in good shape in that department.

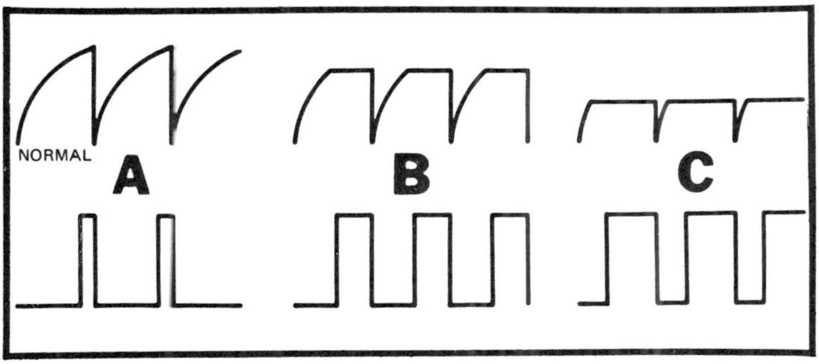

Fig. 2-1. Horizontal output grid waveforms.

This problem doesn't seem to be confined to any particular tube type. I first ran across it a couple of years ago, in a 6JE6. Since then, I've seen it in 6LQ6s, 31LZ6s and so on. So, if you find a creeper, scope the horizontal output tube grid, with the current meter in the cathode, and wait.

This applies to tube *and* hybrid sets, of course. You can see a very similar symptom in solid-state sets. This seems to be due to thermal runaway or its equivalent, excess junction leakage. This is almost always thermal, and the leakage grows progressively worse as the set plays. Read the current drain and scope the base of the output transistor to make sure that the drive pulses are of the correct *width*. They're different in shape to those of tube sets, but the basic principle is still the same. If the pulse width stays the same, try replacing the horizontal output transistor.

HEEDING THE CLUES

Radio-Electronic's service clinic has a fully equipped service lab. Here, tests are run on TV, radio, and audio equipment, and the new test equipment is checked out in actual operation. Test methods can also be checked to see which give the best results in the least time.

Detailed notes are kept on each job, for time-and-motion studies of techniques, equipment, and diagnosis methods. These can be very helpful, and also somewhat embarrassing at times. Here are the actual notes on one job, transcribed verbatim from the clipboard.

Set: RCA KCS-169 hybrid TV.

Symptoms: No high voltage. Sound okay.

Tests: Neon-lamp rf indicator shows what seems to be normal glow on high-voltage rectifier plate lead. No high voltage at all on 1X2 filament. One wire on wrong pin. Corrected. Still no HV.

Question: Are horizontal output tube and flyback putting out enough to light 1X2 filament? No visible glow.

Test: Wind paper clip to make "socket" for No. 222 flashlight lamp. Remove 1X2, tack-solder to filament wires. No glow in pilot lamp.

Test: Set 33GY7 horizontal output tube on cathode-break adapter. Read cathode current (original tube: 150 mA). Schematic says 110 mA.

Try new tube. Same; same current, still no light in 1X2. Try new 1X2 and socket, tacked in. Omit 2.2-ohm resistor. Now read about 2 kV on socket. No light on screen.

Check flyback for shorts, in-circuit. Flyback tester says flyback shorted. Slide chassis out, remove flyback. As anticipated, now tests okay; no shorts. Replace. Disconnect horiz yoke; check. Flyback tester says shorted. Read dc resistance of yoke. Schematic calls for 30 ohms. Yoke reads 30 ohms. Check resistance of each half from centertap; 15 ohms each. Yoke removed from list of suspects.

Second go-round. Check scope waveform loosely coupled to 1X2 plate lead (Fig. 2-2). Scope shows very low peaks (I *knew* that!) and excessive ringing (I didn't know that). Check boost voltage. Reads +165V; B+ reads +135V (normal). Disconnect boost capacitor; checks okay.

Disconnect yoke. Turn set on. High voltage goes up to 5 kV; bright vertical line now appears on screen. Cathode current 33GY7 now drops to 120 mA. Recheck yoke; still shows shorted on flybacker. Disconnect yoke balancing capacitor; checks okay, naturally. Substitute variable inductor for horizontal yoke. Cathode current of 33GY7 now drops to 110 mA; high voltage jumps up to 9 kV, vertical screen line very bright.

This is a very good example of how we can walk right over a key clue without seeing it! Did you? It's easy with our regular 20:20 hindsight, but it was just as obvious when we made the original test! The *test equipment* told me that the yoke was shorted, but I deliberately talked myself out of it and kept on. Right then, I should have stopped and substituted another yoke or the test inductor.

Here are the final lines on the clipboard:

Conclusion: When the flyback tester tells you that the yoke is shorted, *believe it, stupid!* Don't just sit there—*get a new yoke!*

SAME TROUBLE, DIFFERENT CAUSES!

A while back, I was awarded the job of working over a pair of identical TV chassis with the same symptoms. Both had a vertical height of about 2 inches! Plenty of width, etc., but this was a little bit short.

Fig. 2-2. Excessive ringing at 1X2 plate.

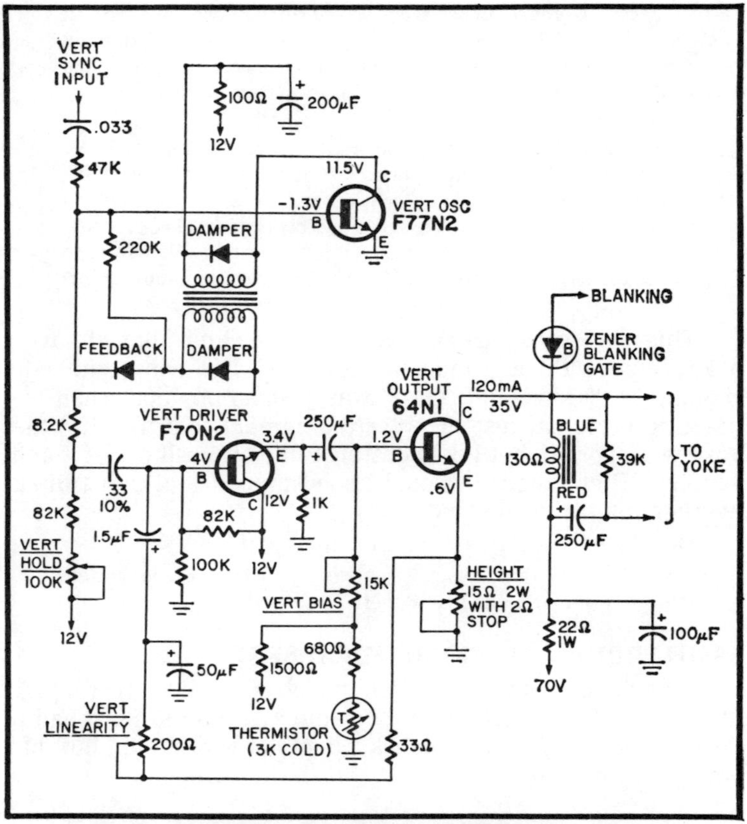

Fig. 2-3. Magnavox 908 vertical circuit.

Figure 2-3 shows the vertical circuit used—both were Magnavox 908 solid-state black & white sets. My scope soon showed me that the drive signal waveform on the base of the 64N1 vertical output transistor wasn't all that it should be!

It was on frequency, all right, at 60 Hz. (In fact, I could see a locked picture on the wee raster, so I knew that already.) But the amplitude left a bit to be desired. It was about 0.2V and the waveform was severely flattened, as you can see in Fig. 2-4.

Something was chopping this thing off right at the knees. A few dc voltage measurements around the driver stage brought me to the possibility of a bad transistor in the vertical driver stage. After some struggling the transistor was pulled and checked. Sure enough. Very bad (leaky).

The original was a type F70N2, an npn silicon. A GE-10 or HEP-53 is a good replacement. So I stuffed the leads through the holes and tacked it in. Cross fingers, apply power.

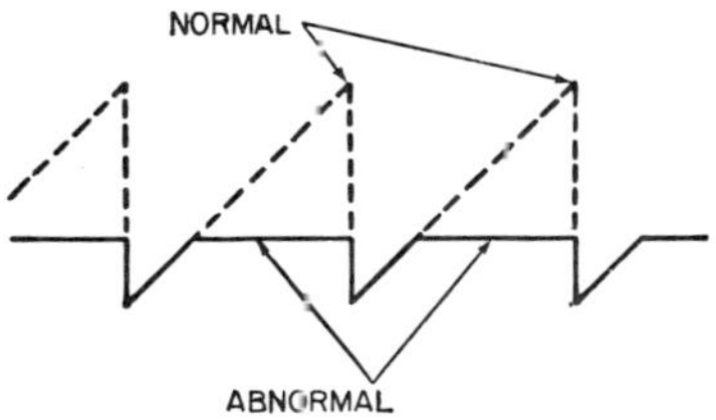

Fig. 2-4. Flattened waveform in Magnavox 908.

Waveform comes up almost instantly, now looking a lot more like "what it says in the book." When the single tube in this set (the picture tube) warmed up, I had a full raster. A few adjustments and a check on the collector current of the vertical output transistor to make sure that it wasn't pulling too much current, and that took care of that.

Now I get out the other one. Same symptoms. In fact, identical; my scope showed exactly the same waveform on the base of the vertical output transistor! Same amplitude, same clipping, everything. "Ho, hum. Gets monotonous, doesn't it?" I say to myself, taking out the vertical driver transistor again.

Things were different now, though. You never saw a better transistor than this one. At least that's what my reliable transistor tester tells me! No leakage, plenty of beta, and everything. Just for luck, I reached for a new replacement transistor and temporarily tacked it into the circuit from the underside.

Firing it up showed that I still had the same chopped-off waveform! No raster, or at least no more than I had to begin with. So I set out after the cause.

A few resistance readings around the driver stage with the transistor out showed something odd. There was not enough resistance to ground from the base. Since this was normally "blocked" by a 0.33 μF coupling capacitor back to the oscillator, and a 1.5 μF capacitor to the vertical linearity control, I should have had 60,000 ohms from the open base terminal (100K to ground, 82K to the +12V source).

Instead, I had about 2000 ohms. Incidentally, base and emitter voltages on this transistor should have been 6.2 and 5.6V, and they were both very low—about 1.0V. (They had been the same on the other one, with a bad transistor!) A little judicious unhooking of things and rechecking resistances soon showed up the culprit—a capacitor!

The 1.5 μF shaping capacitor, from base to vertical linearity control, had about 2000 ohms leakage. Since this went directly to a 50 μF electrolytic, it was shunting the vertical drive signal and causing drive signal clipping and knocking down the bias voltages on the driver transistor, all at the same time. I had a hard time *finding* this capacitor! I'd pull the solder out of the terminals from below, then reach up and feel what I thought was it; however, it wouldn't pull out! I eventually turned the board over, and found that the offending capacitor was one of those tantalum units that looks exactly like a silicon rectifier!

TUBE-TYPE VERTICAL OUTPUT CIRCUITS

There's one important circuit, used in all TV sets, that can be confusing. It's actually very simple, but by the time the manufacturer gets through adding things to it, it doesn't *look* simple. This is the vertical oscillator and output stage. The one I'm referring to is the almost standard multivibrator, using dissimilar triodes such as the 7GF7, and many others. One is a voltage amplifier, the other a power amplifier.

Figure 2-5 shows a simplified version of this—the bare bones of the multivibrator. Put your hand over the parts on top and what do you see? A plain RC coupled audio amplifier. Now, uncover the top parts and you'll see that this is a *feedback* from output to input. In other words, almost 100% regenerative feedback (tending to cause oscillation.)

So if the parts in here (all *nine* of them, plus the tube) are good, it *has* to oscillate! If it doesn't oscillate, one (or more) of the parts is bad. All of them are critical. You can't take any dc leakage through the capacitors, or any drastic change in resistor values. This upsets the frequency. With such a small

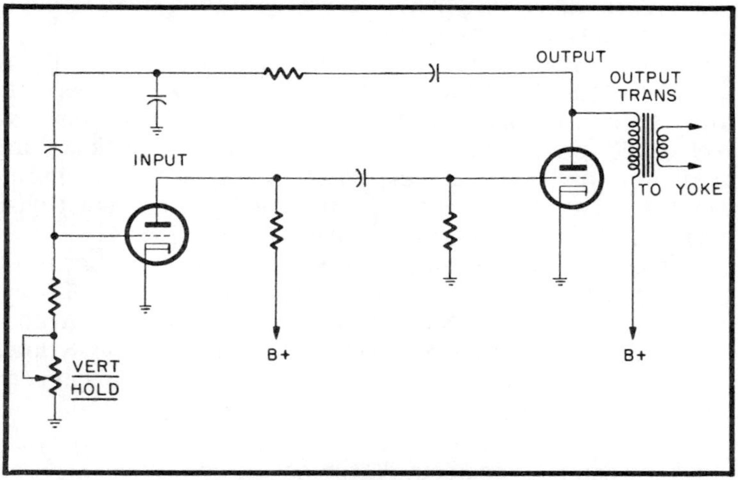

Fig. 2-5. Basic multivibrator.

number of active parts, you can "shotgun" this circuit if you want to—replace *all* the parts at once.

However, before we start loading the shotgun, let's look at some key symptoms. If you know how the thing works, and what it's *supposed* to do, it won't take long to find out if it's doing it. For example: "The picture rolls up or down, but it won't stop!" Do you check the vertical *oscillator*? If you do, you're wasting valuable time. This is a vertical sync problem! All the oscillator is supposed to do is "make the picture roll up or down." In plain words, run slightly faster or slower than the correct frequency. If it does that, forget the oscillator and go for the sync, which is obviously missing.

Okay, what *does* indicate oscillator trouble? When the picture will not reverse direction! It rolls down or up, at all times; or runs so fast that you can see only half a picture, or so slow that you see two complete pictures. Now you've got oscillator trouble. It is always directly related to the *quality* of the parts in the oscillator circuit. A very small leakage in a capacitor in the feedback loop and the oscillator frequency changes. A change in the value of a resistor in the hold control circuit, the plate circuit, or any other also changes the operating frequency, and the hold control won't make it reverse, as it must.

VERTICAL OUTPUT TRANSFORMER

Before we leave this section for the moment, let's look at a part which is often replaced unnecessarily; the vertical output transformer. A careful analysis of the symptoms will show you that in most cases, it couldn't be bad. An output transformer can have two defects; it can be open, in which case you get no vertical sweep at all, and it can be checked with an ohmmeter. Or it can have shorted turns. In this case your linearity is badly upset, usually stretched out near the top of the picture. (Don't replace it yet. Look for a bad electrolytic cathode bypass on the output tube. If it is open, you'll get very much the same symptoms. Bridge a new one across the original to find out.)

YOKE

Yokes are another often replaced part, when they shouldn't be. An open yoke can cause no vertical deflection at all, and can be caught with the ohmmeter. A badly shorted yoke shows you the old keystone raster. The narrow end of the wedge points toward the winding that is shorted.

There is one yoke defect that is quite rare, but possible. If you see a quite severe "bunching" or squeezing up of the

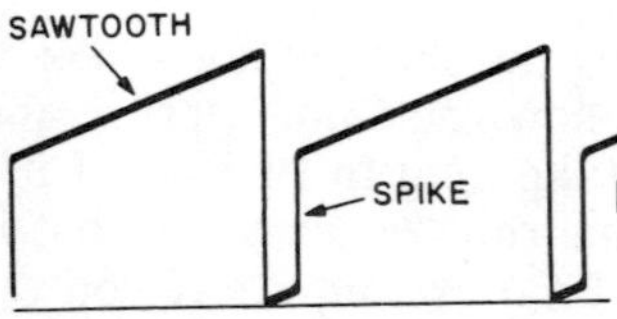

Fig. 2-6. Two-part vertical waveform.

scanning lines, at or near the middle of the raster (not at the very top, or the very bottom), look out. This is sometimes caused by a small short (only one or two turns) in one of the yoke windings. If you have this kind of symptom, and all of the coupling capacitors, bypass capacitors, etc. are good, look for it.

While not universally true, sometimes you can pin it down by checking the dc resistance of the two halves of the vertical yoke. They should match closely. If you find a difference of several ohms, there is a distinct possibility that the squeezing of lines means a shorted yoke winding. Try another yoke of about the same inductance. If you can get a *linear* raster, even though it falls short of filling the screen, a new yoke is indicated.

WHITE VERTICAL LINE

Multivibrators usually produce something like a rectangular waveform. What we need is a specially shaped sawtooth, with a negative-going spike, to get good linear sweep. Figure 2-6 shows a typical waveform.

It actually has two parts; a linear "sawtooth" and a sharp negative-going "spike." How do we get this thing? Simple; by connecting a little RC network from the plate of the input tube to ground, as shown in Fig. 2-7. Each part does its own thing. The capacitor charges to produce the saw, and the resistor affects the spike. We need this negative-going spike to turn off the output tube quickly and completely during vertical retrace. (We also turn off the picture tube with the same spike, if you were wondering.) This is a "saw-forming" network.

This is very important, and the parts are critical. The saw part of the trapezoid must be linear. However, the charging curve of a capacitor is anything but linear. It's exponential, as in Fig. 2-8. So, what do we do? We use only a *small part* of the whole curve, right at the start, where it *is* linear! If we let the capacitor rise to full charge, or rather to 63% of full charge, we'd have "one full time constant" (and a heck of a looking curve).

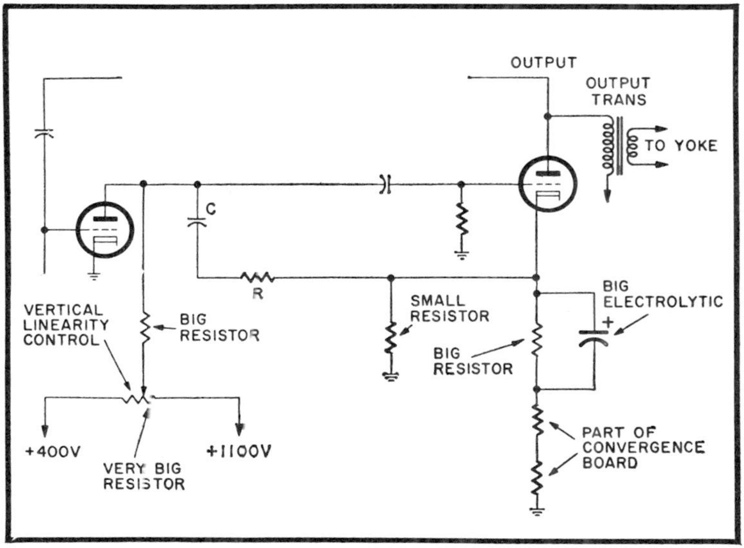

Fig. 2-7. Adding RC network to multivibrator produces waveform.

The time needed for a capacitor to charge up to 63% of the applied voltage, through the resistance in the circuit, is one time-constant. This is figured by multiplying the resistance, in megohms, by the capacitance in microfarads, and you come out with the time in seconds. T = RC. To make sure that this thing uses only the *linear* part of the curve, we make sure that the time constant is quite a good deal *longer* than the operating frequency of the circuit, $^1/_{60}$ second. In the actual set from which this circuit was taken, the capacitor is a 0.039 μF, and the series resistance, all together, is about 4 megohms.

In operation, this capacitor starts to charge when the input tube is cut off (by the feedback pulse coming from the output tube through the network). Its plate voltage starts to rise to the full supply voltage. Before it can get there, or get off the linear part of the curve, the input tube is turned on by the sync; it conducts heavily, and its plate voltage drops practically to

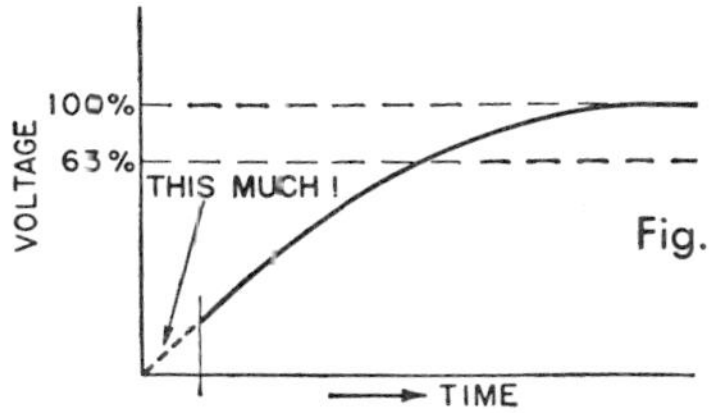

Fig. 2-8. Capacitor charging curve.

zero—the saw-former capacitor discharges. This forms a negative-going spike (not a negative spike; a negative-going spike).

Now, let's add some controls, as in Fig. 2-7. One side of the capacitor discharges to ground, through small resistance R, then to the cathode resistor of the output stage. The *complete* path for this is through the little resistors to ground, to the dc power supply, then back up to the plate side of the capacitor through some very big resistors. (That's where the long time constant comes from.)

Now we come to the first of the "areas of confusion" about this circuit. Notice the high-resistance control, connected between +400V and +1100V, and marked *vertical linearity*. This happens to be in a color set, but they all work the same. Actually, this control regulates the charging time of the capacitor, because it is in series *with the voltage supply*. It is also in the plate circuit of the input tube.

So you'll sometimes find this one marked *vertical height*, or *size*! Same control, same reaction; but different names!

VERTICAL CONTROLS

In the last section, we found a control going under two different names. This time we'll see another one. An obviously suspicious pair of characters! Figure 2-9 shows more of the same circuit, with the other control, and the rest of the little gubbins added. You will find minor modifications of this (a Magnavox 921, 933, etc.) but they'll all be recognizable, and they all work on the same principle.

Vertical Height (Size Control)

The new control is connected to the grid of the output stage. Through voltage divider action from B+ (405V), it adjusts the bias on the grid and so controls the gain of this tube. This in turn controls the height of the raster. So this one is found under the name *vertical height control*.

In normal operation this grid has a pulse of at least 100V p-p on it. So it develops a good-sized negative bias by grid-leak action. This tends to cut down on the output. Some positive voltage is added through the height control to cancel a little of this negative voltage and keep the tube working at high output.

False Names and Adjustments

The *vertical linearity control* varies the charging time of the 0.039 μF capacitor, the sawtooth waveform generator. It does control the linearity. Of course, it also varies the plate voltage of the input section. It also affects the vertical size

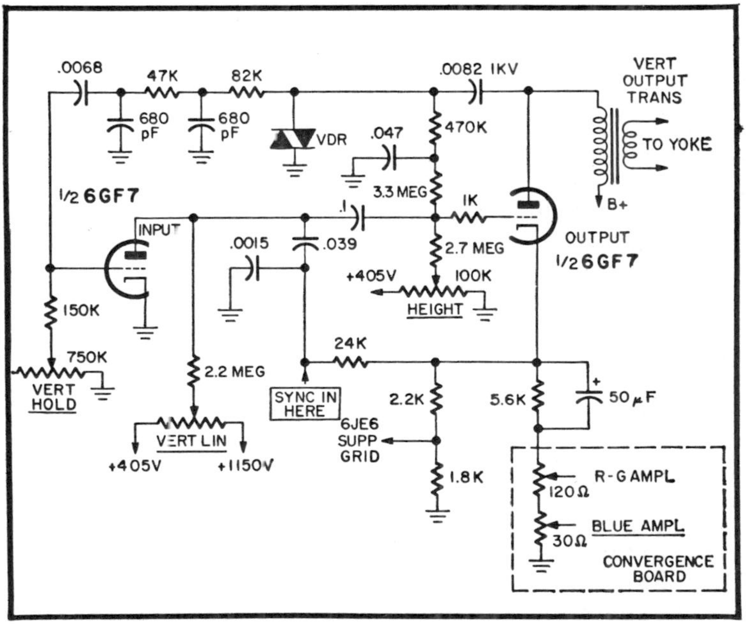

Fig. 2-9. Vertical circuit with voltage-dependent resistor as part of control element.

(height). That's why you find this control marked *vertical size* in some sets.

The *vertical height* control also has quite an effect on vertical linearity! So in some sets, you'll see this one marked *vertical linearity*! Don't let this confuse you. As long as you've got one of each, go right ahead. Actually, you will have to adjust both of them to get the raster to the necessary "full and linear" condition.

Feedback Loop

Across the top of Fig. 2-9 you see what looks like a very complex circuit: the feedback loop that makes the circuit oscillate. It picks up a high-voltage pulse, 800 to 1000V p-p at the plate of the output section, cuts it down and shapes it up a little, then feeds it back to the grid of the input section. Its basic action is exactly like that of a vertical integrator.

The Mysterious VDR

There is one new part in there. It is found in a lot of late-model sets, especially the wide-angle deflection types. It is the *vdr*, or voltage-dependent resistor. This special resistor is made of a semiconductor material that has the odd property of *changing* its resistance in relation to the voltage across it. Unlike the thermistor (which goes down in resistance as the *current* through it increases), the reaction of the vdr is very fast. It can change resistance in a fraction of a second. The *higher* the voltage, the *lower* the resistance. This reaction is used as a regulator in many circuits.

In this one, during retrace time the high-voltage pulse shows up on the plate side of the 8200 pF capacitor. The vdr is shunted across the other side of this capacitor. When the high voltage appears, its resistance goes down; so the capacitor charges to a high voltage. The pulse, during this time, is fed through the capacitor and on to the other parts in the feedback loop.

After the pulse passes, there is a sweep-time period when the voltage on the grid of the output tube must rise slowly in a very linear way (the sawtooth part). The capacitor discharges slowly through those two big resistors (470K and 3.3M) to the output stage grid. This negative voltage also helps cancel some of that positive voltage being fed in by the height control. The result of all these actions is to improve vertical linearity.

It also regulates the output. If the line voltage rises, the B+ goes up with it, and the positive spike gets bigger. The raster would be overscanned if not controlled. The higher voltage of the spike, though, *reduces* the vdr value, and the 8200 pF capacitor charges to a *more negative* value! This, applied to the grid of the output stage, holds down the output, and the raster stays the same size.

Testing the VDR

Because of its peculiar characteristics, the vdr is hard to test with ohmmeters, etc. You can make a test setup, with a high-voltage supply, dc milliammeter, and so on, but substitution is a lot faster.

If you suspect the vdr is causing trouble, which is rare, but always possible, there's a rough check you can make. The only problem would be a shorted vdr. If it opens, you'll have overscan, and not so much regulation. Disconnect one end of the vdr and tack a 1M resistor in its place. If the vdr was defective, the raster will fill out, and you may see some overscan. No regulation, of course. Replace with an exact replacement.

Vertical Sync

In the first part of this section we mentioned sync troubles. In the circuit (Fig. 2-9), the vertical sync is fed to the bottom end of the saw-former capacitor (0.039 μF) as a negative-going pulse. It is then fed on through the 0.1 μF coupling capacitor to the grid of the output tube. If the *vertical hold* control is properly set, the oscillator will be running just a little bit *below* the correct frequency, and the sync pulses will cause the oscillator to "fire" at exactly the right time.

The most important thing about the vertical sync is its *amplitude*. The sync pulses must be high enough to fire the oscillator or you'll have a condition of weak sync-lock, jitter, etc. This kind of problem is usually caused by a weak sync separator tube, low dc voltages on that stage and leaky vertical integrators.

Weak-Sync Problems

In some cases you'll see what looks for all the world like weak sync. In fact, the vertical lock will be pretty bad, although you'll see the characteristic jerk or snap of the picture as the blanking bar rolls past the bottom of the screen. This can be and often is due to misadjusted vertical height and linearity controls!

If they are set so the raster is considerably overscanned, they will cause flattening of the oscillator waveform at the point where it should be sharply peaked (where the sync fires it). The waveform won't approach the firing point as rapidly as it should. The sync can then make it fire early, or late, and the whole thing looks like poor sync. You can cause this condition deliberately by overscanning the raster and checking the vertical hold action. For best sync, set the controls so the raster is overscanned by not more than half an inch at the top and bottom.

Other Troubles

A slight leakage through the 0.1 μF coupling capacitor also causes a similar symptom. If this leakage gets worse, it flattens the peak of the waveform and causes foldover at the bottom of the screen. (Since the beam sweeps from top to bottom, remember that the *top* of the waveform represents the *bottom* of the TV screen!)

The best way to isolate sync trouble is to kill the vertical oscillator by pulling the tube, grounding a grid, etc. Then check the sync amplitude at the sync-injection point. Watch for two things; loss of amplitude and any sign of video in the vertical sync. The last can be due to incorrect operating point in the sync separator, a defective integrator, etc. It will cause vertical jitter, due to the varying voltages of the video. Vertical sync should always be "clean."

Oddball Troubles

In color sets you can get some very odd vertical troubles due to the many uses we make of the vertical stage. As you can see in Fig. 2-9, we are taking off a dc voltage of about +30V from the 1800-ohm resistor in the cathode. This is fed to the suppressor grid of the horizontal output tube to prevent snivets on UHF. It has nothing to do with the vertical output stage; it's just a handy source of 30V dc. However, if the 6JE6 tube has a bad internal short, the 1800-ohm resistor burns up and all of a sudden we've got vertical output troubles.

Actually, because of the peculiarities of this circuit, this won't kill the vertical sweep completely as you'd expect with that cathode open. It does not open completely. You always have a parallel path to ground through the convergence board! Instead of losing all the sweep, you lose about half the height. People get very flatheaded and nothing you can do will help in the vertical controls.

When you run into this kind of problem, check the cathode circuit of the output stage. One key clue is the cathode voltage. Because of the higher resistance now present, this will be well above normal, usually from 75 to 100% more. If either the 2200- or 1800-ohm resistors open you'll have the 5600-ohm resistor and those two small controls left in the dc path to ground. This increased resistance raises the cathode voltage and the increased negative bias on the grid cuts the gain of the stage. (Positive voltage on a cathode is the same as the negative voltage on a grid.)

If that 50 μF electrolytic capacitor across the 5600-ohm resistor opens, you'll see very little effect in the vertical size and linearity, but your vertical convergence will go way off. This is a coupling capacitor, feeding convergence waveforms to the convergence board.

If the 50 μF capacitor shorts, you'll upset the total resistance in the cathode circuit. All you'll have left in the higher-resistance branch will be those two little controls on the convergence board, usually about 100 ohms or so. The bias voltage will drop and the picture will usually stretch at the top beyond range of the controls. Once again, a simple resistance test from output cathode to ground will catch this.

Broken wires in the cable from chassis to convergence board; wires in this cable pinched under clamps or shorted to ground; open controls on the convergence board—all can affect your vertical output stage and sweep. If you find mysterious cases of vertical troubles, don't forget to check out that cathode circuit of the vertical output tube!

As a word of comfort before we move along, the majority of your troubles will be easy ones—bad tubes (or transistors) leaky capacitors, burnt resistors, and so on. As long as you *know* how the thing works and how they hooked it up this time, you'll be okay. Part values will be changed between this and others, of course, but the purpose will still be the same.

COMPLEMENTARY VERTICAL SWEEP CIRCUITS

Looking into some of the newer solid-state TV sets, color or black & white, you might say, "A component is missing!" It is! It's the vertical output transformer. Look at Fig. 2-10. What's this? The guys who say "an output transformerless audio stage" are right; the ones who said "a vertical output stage" are also right. If L is a speaker, audio; if L is the vertical windings of the deflection yoke, it's a vertical output stage. You'll find this in such new sets as GE's JA and MA color chassis, and others.

The theory is the same as in the audio applications. This is a *complementary-symmetry* output stage, identical to those used in audio amplifiers. The output is taken of at the midpoint—in this case, the common-emitter connections (which can lead to some dandy new symptoms that we'll examine shortly).

The only difference between this and an audio amplifier is the signal input. Instead of rock music, we are dealing with a sawtooth signal developed by an oscillator. The phase inversion needed for push-pull output is handled in a differential-pair stage and a driver.

Some circuits use a buffer amplifier between the oscillator stage and the diff-amp stage. The output circuit here is a true

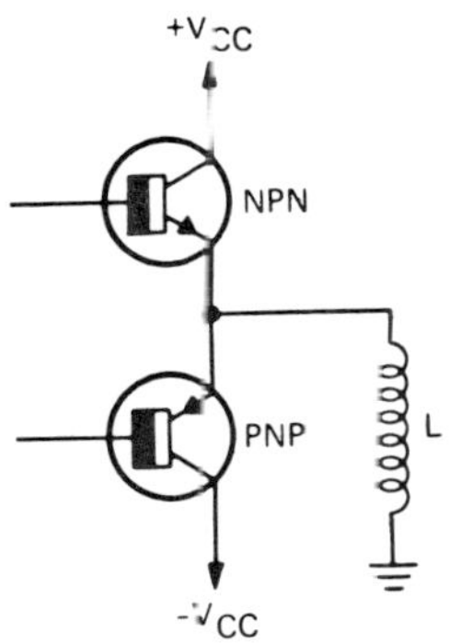

Fig. 2-10. Transformerless output stage.

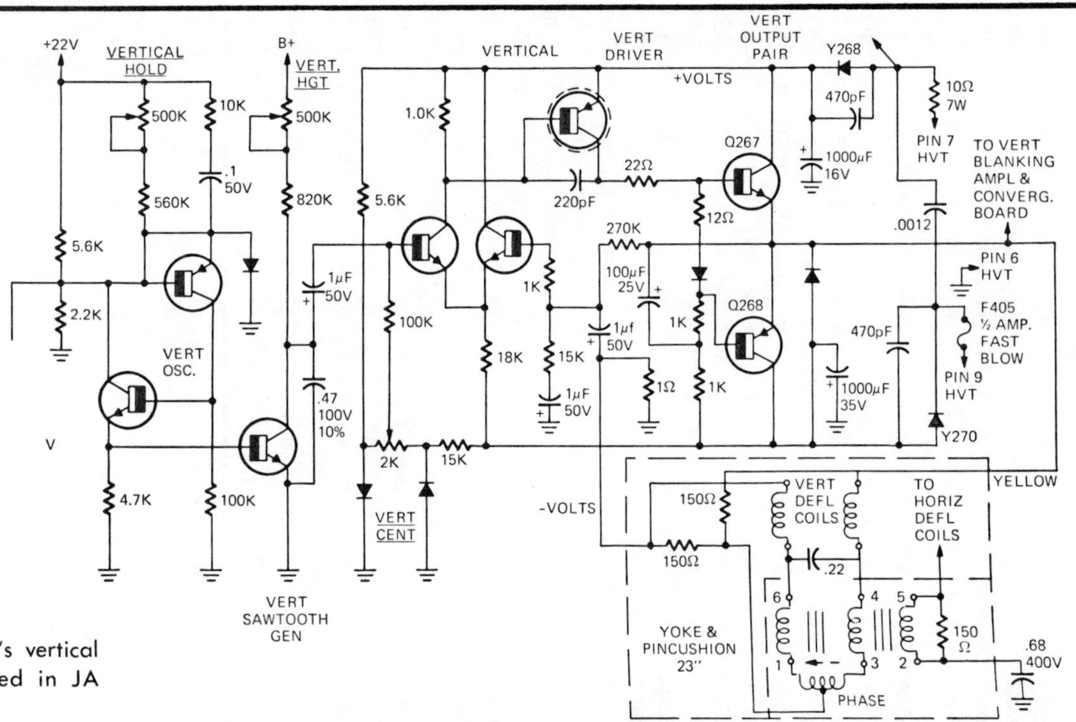

Fig. 2-11. GE's vertical circuit, as used in JA chassis.

complementary-symmetry type, with opposite polarity voltage supplies for the npn and pnp output transistors. So the big capacitor used in single-ended audio circuits like this isn't needed. The yoke returns to ground, usually through the pincushion corrector and a feedback network.

In the GE circuits, this feedback goes to the differential-amplifier stages, and is used to correct for any distortion in the output. A perfect sawtooth is developed, and there goes another old familiar component—the vertical linearity control. It isn't needed in this kind of circuit (and it won't be missed). Figure 2-11 shows the complete circuit, as used in the GE JA chassis.

Along with the new circuitry, you will see some "different" symptoms, as compared to our old "stock" vertical symptoms. For example, if you get a set with a nice linear picture, but only on the top half of the raster, what would you say? The picture isn't compressed; half of it's not there!

This is a real dandy and it can lead us straight to the trouble. (And we love them, don't we?) If the top half of the picture is missing, Q268 (Fig. 2-11) is open. If the bottom half is gone, Q267 is open.

The *vertical centering* control voltage is fed to the differential-amplifier stage. So if the picture will not center properly, go straight to this stage and make sure that it will "balance"; both transistors good, etc. There is another handy trick you can use, on certain chassis. (These are the ones where the vertical output transistors are not on the vertical module itself, but on the chassis.) If you have the old familiar symptom of "thin horizontal line—no vertical sweep," try moving the vertical centering control. If the line moves up and down about the right amount, the vertical *output* transistors are good. Go and look in the vertical *oscillator* stage for the cause of the trouble.

In the JA chassis circuit shown, observe the novel method of supplying power to the vertical output transistors, driver, and diff-amps. Note the arrows saying "To pin 7 HVT," "pin 6 HVT" and "pin 9 HVT" (high voltage transformer). These go to a winding on the flyback, which supplies 60V p-p pulses, of opposite polarities, to rectifier diodes Y268 and Y270. Pin 6 is the centertap on the winding. The rectified pulses are filtered by the two big capacitors and provide the dc power needed for the vertical output stage. This has one benefit: if the horizontal output stage fails, you do not have the very bright, thin horizontal line which might burn the screen of the picture tube. When the high voltage goes out, the vertical sweep goes with it.

In the next version of this chassis, the later MA, the dc voltage for the vertical output comes from the low-voltage power supply. A + 13V supply feeds the pnp. A +130V supply feeds the vertical buffer transistor through the *size* control (and *service* switch), and a +23V supply feeds the oscillator.

For a final "new symptom" in the JA chassis, if you see no video and no raster, and fuse F405 is open, check the vertical output transistors; it may be that both are shorted.

READER QUESTIONS

Not Enough Height

This one sounds easy, but somewhere along the line I'm missing something! This Zenith 20Y1C48 came in with no vertical sweep. The 6HE5 vertical output tube was cracked, a 470-ohm resistor was burnt, and the 100 μF electrolytic in the cathode circuit had blown up.

I replaced all these parts, and got a good picture, color and all, but now I don't have enough height; all I can get is about ⅔ of a normal raster. What did I overlook?—H.E., PA.

Same thing I overlook at too-regular intervals! The rest of that vertical-output cathode circuit. In a color set, this has "more to it than meets the eye!" Beside the regular bias resistor and bypass capacitor, the cathode circuit goes sneaking off over to the convergence board, and goes to ground there through several low-resistance controls, etc.! So, your cathode circuit is actually a complex series–parallel network!

If you'll check the cathode *voltage* on the 6HE5, I expect you will find it high. Excess voltage at this point indicates *too much resistance*—in other words, something is open over on the convergence board!

Vertical Linearity Problem

I've two or three problems in this Zenith 24NC31Z! It has a very odd vertical linearity problem, and those resistors across the degaussing coil are burning up! There seems to be a short somewhere, but I can't find it.—O.M., Brooklyn, NY.

Step 1: replace those "resistors" across the degaussing coil. These are not standard resistors, but a vdr (voltage dependent resistor) and a thermistor. There is a good chance that this set has been visited by lightning not too long ago! If so, these parts seem to be the ones most frequently damaged.

If these are bad, you can get some weird effects in your power supply, because of the distortion of the ripple waveform. Replace the vdr/thermistor combination, and see if your other trouble doesn't clear up.

Triple Trouble

Since this happened to me, no initials! This CTC 24 chassis came in from another shop, after having a flyback replaced. Made a picture, but had a very distinct *drive* line right in the middle of the raster.

After some checking, I found the lead to terminal 1 of the flyback hanging open! Aha! Hook this up, and what happens? No raster. Cathode current of the 6JE6 jumps to 400 mA (it was slightly low before) and the high voltage goes down. Eventually, I found that the (original) focus transformer was shorted. Replaced this and things looked much better.

Second problem: After this, the cathode current of the 6JE6 was okay, but the horizontal efficiency coil refused to dip, and the coil ran very hot. Replacing this helped. Now it would dip, and ran much cooler. Evidently, it too had a small (but sufficient!) short. Now things were better except for a tendency of the picture to jitter and flicker. Looked like an intermittent arcing somewhere in the high-voltage circuit.

The high-voltage rectifier socket and leads were checked. All okay, of course. So I checked the focus rectifier by replacement. No help. Scope waveforms around the B++ circuit showed a pronounced jitter. B+ boost was very low and unstable. In fact, it read only about 900V instead of the normal 1100V, although the B+ voltage was up to normal.

Taking the boost-boost rectifier out, I found that I could get a reading on an ohmmeter! Forward resistance was only about 2–3 megohms! A new one read a very high resistance in the same direction, and the back resistance was absolutely infinite even on the 1000-megohm scale. This cured the problem. Evidently, this rectifier was arcing internally, and had partially shorted.

Horizontal Instability

The horizontal oscillator in this Zenith 25MC33 chassis is very unstable. Horizontal hold range is very small. Blanking bar sometimes locks in the center of the screen. I hear a frying in the flyback at times. Vertical sync very good.—J.B., Elmhurst, NY.

Short out the afc. If this lets the horizontal oscillator run on-frequency (that is, make a single, floating picture), the oscillator itself is okay. Since you have good vertical sync, this shouldn't be in the sync separator. The most likely cause here would be a defective dual diode in the horizontal afc. Try a new one; this is the quickest way.

Horizontal Shrinkage

I've got a peculiar problem! One of my best customers brought in a little Emerson transistor portable TV, a 120771 chassis. It's got a "thermal shrink" of the raster, just like a weak horizontal output tube! The only trouble is, no tube! All solid-state!

This is the one which uses a "gate turnoff" horizontal output device. About 15 minutes after it's turned on, the raster shrinks in about an inch from each side. I've read the current to the GTO, and it's quite a bit high. In other words, the thing is taking too much current, but the narrow raster shows that the yoke isn't getting enough current. I'm baffled! What the heck is this?—S.P., NY.

If your drive signal shows okay, and voltages are up to normal in the first few minutes of operation, I believe you'll find that this is due to a *leaky* gate turnoff switch. Due to the circuit connection, this is effectively in shunt with the yoke. So you're trying to draw normal yoke current, plus the leakage current, and you wind up with a narrow raster.

The exact replacement for this is Emerson part No. 815157. However, a good high-voltage, high-current silicon power transistor in a TO-3 case will replace it! (A power transistor, after all, *is* a turnoff switch.) Delco DS-502 or Motorola HEP-740 can be used as replacements; check the current to make sure 680 mA is normal.

Scanning Lines Squeezed

I've got an odd problem. In a Zenith 14Y33 chassis, the raster is squeezed up about 2 inches from the bottom (covers full screen). This makes a bright line across the picture and distorts it badly. The middle is stretched quite a bit. I've checked all of the parts in the feedback loop, since the vertical sync isn't too good. They are all okay; tubes replaced.

I suspect the vertical output transformer. What do you think?—L.C., Hatfield, AR.

The first time I saw this, I suspected the transformer, too. However, when I replaced it and it didn't help at all, I began to doubt the validity of this diagnosis! This type of vertical problem is almost always due to trouble in the vertical yoke. This has been verified in at least three recent cases.

This will not cause the familiar keystone effect of a heavy short; it is apparently due to a short in the vertical yoke, but not as bad as if it did keystone. On one of these, I checked the resistance of each half, and found one half reading 8 ohms, the other 12. (Normal total, 24 ohms, so each half should have read 12 ohms.) The actual resistance is not so important as the fact that the two halves balance. Should read exactly the same.

Incidentally, vertical-output transformer shorts will tend to make the raster either squeeze or stretch at the top of the picture.

Horizontal Output Transistor

I have a small GE transistor TV with the horizontal output transistor blown. Can't find a replacement that will fit; they're all too small! It's a TB chassis.—J.B., Indianapolis, IN.

I think you'll find that an RCA SK-3035 will replace the GE 15X5 transistor in the output stage. A Motorola HEP-235 for the output or Motorola's latest horizontal deflection output transistor, the HEP-740. This thing has a peak voltage of 750V, so it should hold up.

For the horizontal driver, RCA SK-3020 or HEP 736. This is a small transistor, and a lot of standard "general-purpose audio" transistors will work here.

Be sure to connect a dc milliammeter in series with the horizontal output before you turn it on. Then watch the current. Normal is about 800 mA. If you get a lot more than that, turn it off quick!

No Vertical Sweep

I've got no vertical sweep on this little transistor portable, an Emerson 120771 chassis. All of the transistors check good with my in-circuit tester. I get no sawtooth waveforms at any point in the vertical circuit. I do see "spikes" which are at the vertical frequency, and can be controlled by the vertical hold! What's going on here?—L.S., NY.

If you have "spikes" (which are probably the *amplified* vertical sync pulses going through the circuit), but no "saws," then the trouble is most apt to be in the "saw-forming" circuit. In this type of multivibrator circuit, the sawtooth waveform needed for operation is developed by the slow charging and rapid discharging of a "capacitor"; in quotes because this is actually two separate capacitors, the 25 and 40 μF units between the collector and base of the vertical oscillator transistor. The two capacitors are used with the vertical linearity control connected to the junction to make the sweep linear. The diagram in Fig. 2-12 shows a partial circuit.

So on the face of the evidence, if you have spikes, this means that the amplifier function of the circuit is okay. If you are not developing the sawtooth waveform, one of these

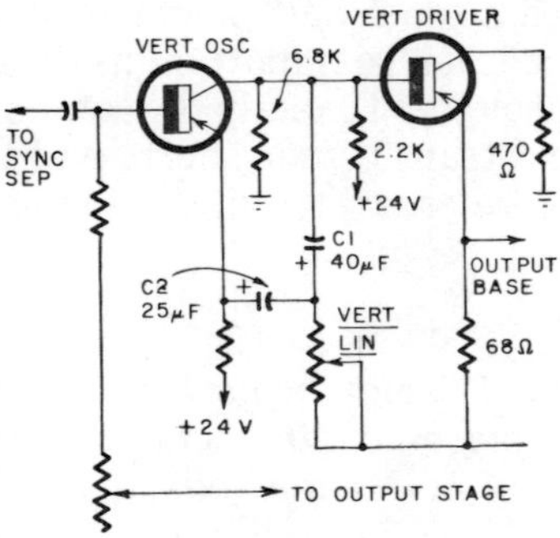

Fig. 2-12. In this partial vertical circuit, C1 and C2 are shown as two capacitors, though they're housed together. Absence of a sawtooth waveform is indicative of a leaky capacitor. Be sure to replace it with an exact-value, close-tolerance type.

capacitors is probably leaky or open. Your dc voltages will probably be very close to normal.

By the way, do not use ordinary aluminum electrolytic capacitors to replace these! Your tolerance will not be tight enough. The originals are the tiny tantalum types; in this timing circuit, you must get very close to the proper values or the circuit will not work.

Repeated Burnout Horizontal Driver Transistor

I have a Philco Q1054BK solid-state TV. Came in with the horizontal driver transistor burnt up. I replaced it, and the new one promptly blew too.—J.G., NJ.

Check the driver transformer, T6. If it is bad, it probably caused the original transistor to blow, and will blow the new ones.

To be safe on things like this, use a Variac in the ac line input. Connect current meters into the collector (or supply) circuit of any suspected transistors, and then bring the line voltage up very slowly. If you see more than normal current being drawn, but the line voltage is only up to about 50–60V, look out! This will save a lot of transistors!

Zenith Color: Vertical Shading

I've a problem with a Zenith 20X1C38 chassis. There's vertical shading on the raster. (See diagram, Fig. 2-13.) It gets more pronounced as contrast is turned up. Color is fine. What is this?—B.M., New Haven, WV.

Probably something in the horizontal blanking diode. Early production of this chassis had two diodes, later runs had three (one of them is a clamp). Check; the blanking is applied

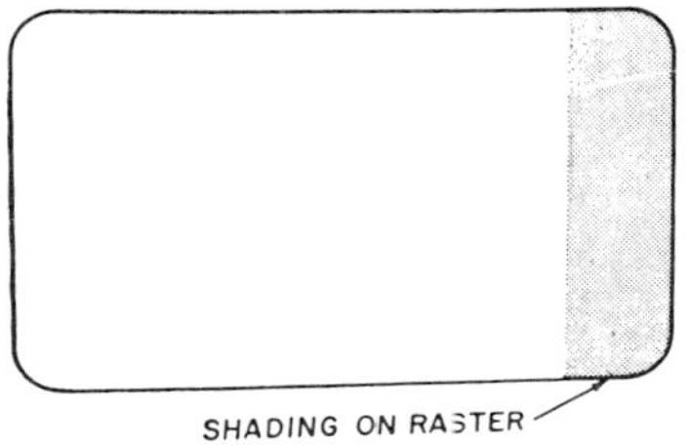

Fig. 2-13. Zenith chassis showed vertical shading at right edge of raster.

in the video amplifiers, through the diodes to the emitter of the second video amplifier, a transistor (see Fig. 2-14). Turning the color off should have no effect, but as you said, turning the contrast down makes it "weaker."

A scope will show you this shading as a "slant" of the video signal, at horizontal-scan rate. However, it's just as easy to pull the horizontal blanking diode and check it for leakage with a transistor tester. Or replace it. Leakage in the video amplifier transistor can cause this, in some cases.

Ripple In Lines, Top and Bottom

In a Heathkit GR-25 color TV, C704 shorted, burned up and almost disintegrated. (See schematic, Fig. 2-15.) I replaced both it and C705 with 500 pF capacitors. Now I have a peculiar rippling in horizontal lines at the top and bottom of the raster—about five or six wiggles all the way across. Lines in the middle of the screen straight.—L.H., Reston, VA.

Let's polish up the crystal ball and have a look! It is quite likely, from your description, that the heart of the trouble is in the pincushion-correction circuitry, since these have the

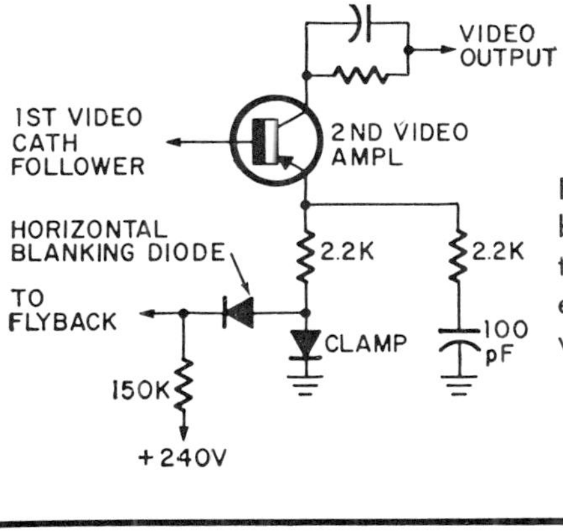

Fig. 2-14. In this circuit, blanking is applied through the diode to the emitter of the second video amplifier.

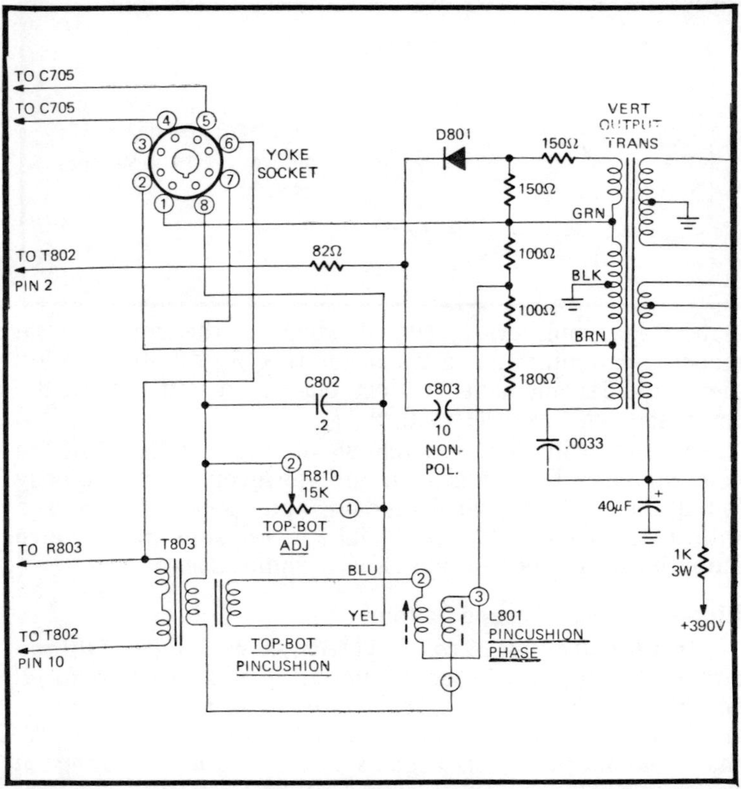

Fig. 2-15. Rippling in this Heathkit GR-25 color set could be attributable to a bad component. Most likely candidates are D801, T803, R810. If these check out okay, the trouble source is probably a bypass or waveshaping capacitor.

maximum effect at the top and bottom of the screen. When C704 (not shown) shorted in the boost circuit it could have blown something in the corrector circuit.

Possibilities: diode D801; pincushion-correction transformer T803, control R810; or caused by the high spike voltage that was undoubtedly fed into this circuit, some of the bypass and shaping capacitors—C803, C802, etc.

As in most color sets, there is a horizontal-frequency spike waveform used for blanking. It has quite a good bit of ripple along the baseline. If something is upset in the circuit and this baseline ripple is allowed to get into places where it shouldn't, it causes trouble. Suggestion: try the pincushion corrector adjustments, and see if they *work* normally. If they don't, then check this circuit for other damaged parts.

False Keystone

I have a pretty distinct keystone raster problem in a Zenith 20Y1C50 color chassis. Still, I've changed the yoke, with no results! I thought that bad yokes were the only thing that caused keystoning!—P.S., Chicago, IL.

In Zenith rectangular-tube chassis with pincushion correction there is a waveshaping capacitor in the screen-grid circuit of the vertical output tube. (See schematic in Fig. 2-16.) It is actually a part of the pincushion-correction circuit. If it opens, you'll get a false keystone symptom, with the raster wider at the top than the bottom, by about 1.0 to 1.5 inch. It may be an 18 or 30 μF capacitor, depending on the chassis number.

Horizontal Drift

I'm working on a Panasonic AN-32 portable TV. It works fine for a while, then the picture starts slipping horizontally after warmup. I suspected the sync tube, although I had an awful time trying to find one. Didn't help. What could cause this?—R.S., Endicott, NY.

Lots of things. Try a new set of afc diodes first. Sync tubes seldom cause *horizontal* sync problems, unless they're completely dead. (Horizontal sync works on *phase*, vertical sync on *amplitude*. So any trouble in the sync tube usually shows up as loss of vertical sync first.)

One more: In this circuit there is a 5.6-megohm resistor from the horizontal oscillator grid to 120V. This is a "stabilizer" or holddown resistor, intended to keep the grid

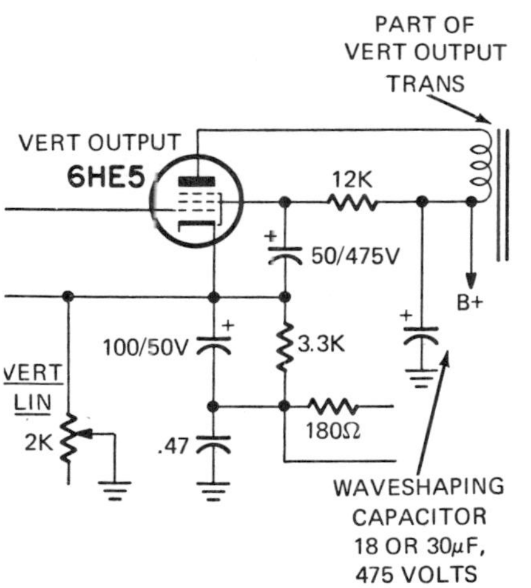

Fig. 2-16. If the waveshaping capacitor opens, the result will be a false keystone symptom.

from being driven too far negative. If this changes value, it will change the oscillator frequency. Check it for heat sensitivity by touching it with a hot soldering iron.

Horizontal Trouble, Magnavox 908

I've horizontal drive trouble with a Magnavox 908 transistor TV. I get about 7—8 kV of high voltage, but the base waveform on the horizontal output transistor is very distorted. Not as big as it should be, either. Tried a new horizontal output transistor, but that didn't help. I get a peculiar waveform on the primary of the horizontal driver transformer, too.—R.G., Salina, KS.

This can be annoying! Check the dc voltages on the horizontal *driver* transistor. I have a deep suspicion that you'll find almost the same voltage on both collector and base! This transistor is probably *shorted.*

The question immediately arises, "How can we get any signal at all through it if it's shorted?" You have apparently got a base—collector short and with the circuit connections used here, you *can* get some drive signal to flow through and get to the horizontal output!

The two isolated transformer windings let this happen. In direct coupled circuits no, but in this case it's possible.

Vertical Retrace

I'm getting too much vertical-retrace line pattern on a Zenith 20Y1C38. With the brightness turned up quite a bit, it isn't too bad, but when the brightness is down to normal it is bad. Can't find it.—A.T., Schenectady, NY.

The vertical blanking in this chassis goes to the *base* of the second video amplifier transistor. This is a busy little rascal; vertical blanking to its base, horizontal blanking to its emitter, and video straight through!

There's an RC network in the base circuit, going to the cathode of the 6HE5 vertical output tube (see Fig. 2-17). Check these out; open C66 or leaky C67 will reduce the amplitude of the vertical blanking pulse. Also, check the second video transistor for leakage.

Vertical Deflection With Tube Out

This RCA KCS-136M has me mystified. The vertical linearity control burns up as fast as I put a new one in. The width control was shorted so I replaced that too, and the linearity control still burns up. Another wild one. With the tube out, I still get about 2 inches of vertical deflection.—L.H., Hitchcock, TX.

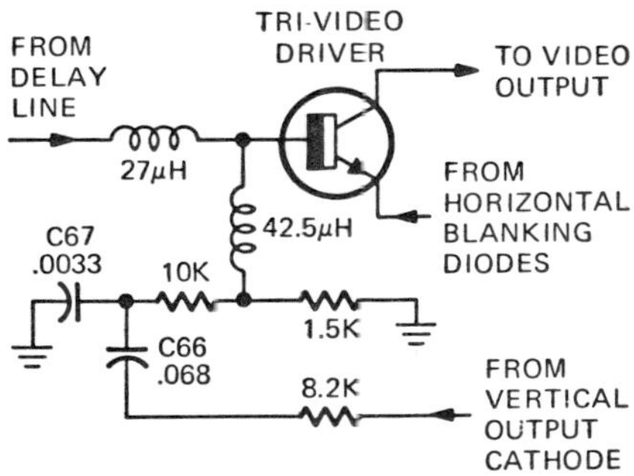

Fig. 2-17. This stage is a solid-state portion of a hybrid chassis. Capacitors C66 and C67 could cause trouble if they develop leakage.

In this chassis (Fig. 2-18), if the width coil shorts between the two windings, it *will* put the boost voltage on the cathode of the vertical output tube. This will burn up the little 2-watt 850-ohm vertical linearity control. However, this control is also the cathode resistor for the 6EM7 triode.

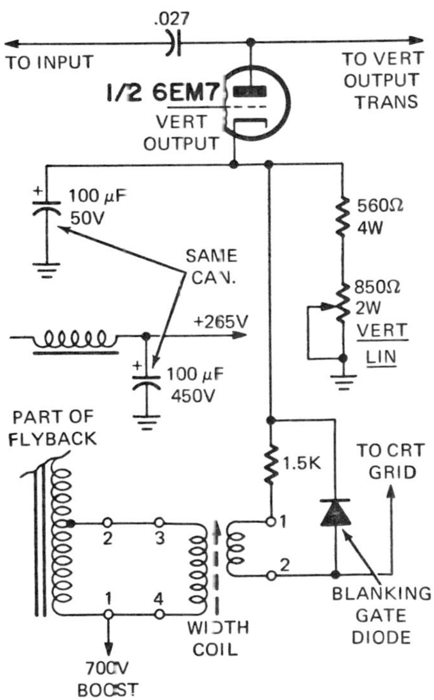

Fig. 2-18. In this circuit, a short in the width coil puts B+ boost on the cathode of the vertical output tube. The inevitable result is a burned-up linearity control.

So if something is upsetting the bias on this tube, the control will burn up because of the excess current being drawn. Check the bias on the grid. Also pull the tube and check the cathode to see *what's* there. Check for dc voltage, which should be zero, and for any sign of hum signal with a scope. This should also be zero.

I suspect that deflection without the tube is caused by *hum* getting into the vertical yoke. The scope will tell you for sure. There is a 100 μF electrolytic on the cathode. This should stop *anything* in the way of hum or ripple, if it's not open. Second possibility; rare, but has happened—an intersection short between this 100 μF section and the 100 μF section in the multiple electrolytic. This would put the +265V supply on the 6EM7 cathode, and again the linearity control would burn up.

Trap Setting, and So On

I have problems with setting traps in sweep alignment. Should you set the marker as low as possible, or what? How do you do this properly?—D.M., Camdenton, MO.

Any i-f trap should be set for *minimum* response at that frequency. For example, the 41.25 MHz sound trap, if it is

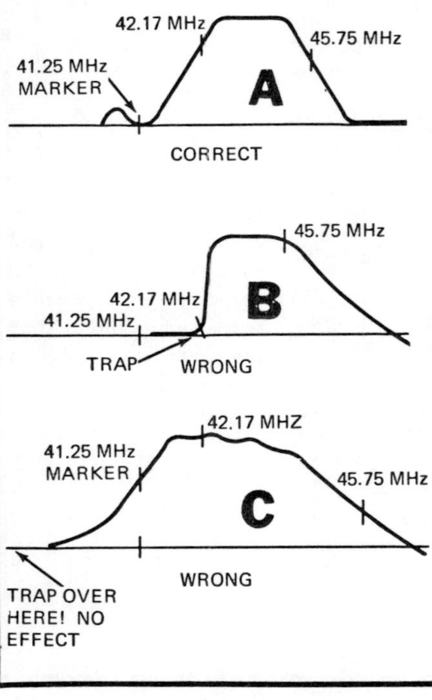

Fig. 2-19. A trap should always be set for minimum response at its frequency of operation. The correct waveform null is shown in A; in B, the trap frequency cuts into the response curve; and in C, the trap is so far off frequency that it doesn't show up at all.

working properly, will make a "notch" in the response curve. The diagram of Fig. 2-19 shows several possible conditions and the correct setting.

Moving the trap tuning core should make this notch "walk" back and forth along the curve. If you can't find the notch, tune the trap all the way from one end to the other. You may find it, as I did once, clear over on the *wrong* side of the curve! Set it so that your marker is exactly in the middle of the notch.

Loss of Height

The raster on this Zenith portable is only about 2 inches high, and ripply. Changed the tube, changed a vertical linearity control which was open, and so on. Now everything checks fine but it won't work!—H.M., Gainsville, FL.

Under the circumstances I'd check the *yoke*. We have had a good deal of assorted yoke troubles lately, in several makes. Mainly in the little import yokes with blue plastic.

Open the yoke circuit, find the centertap, and check the two halves for *balance*. While the resistance won't tell you too much, it must be the same for each half. Any unbalance is a sign of trouble. Look for signs of arcing, etc. Best check: connect a new yoke.

No Vertical Sweep

This Zenith 24NC31Z came in with no raster or sound. I found a bad resistor in the B+, changed it, and it worked. When I took it back, all I could get in the home was a horizontal bar about an inch high! Checked the service switch, tube, and so on. No help.—A.S., St. Paul, MN.

I did the same thing not long ago! Set working perfectly on bench, on extension cables; back in cabinet, no vertical sweep. So do what I did—check to see what isn't hooked up! There has to be something.

Crystal-ball diagnosis: check the convergence yoke, plug, and socket. If this isn't plugged in, or has a bad contact on one certain pin, you'll get exactly this symptom. The vertical output tube cathode circuit returns to ground *through* the convergence circuitry. So, without this plug in properly, you have far too much resistance in this cathode, and lose practically all your height due to excessive bias.

Horizontal Oscillator Setup

There are four or five pictures across the screen of this RCA KCS-130 chassis. I've changed the oscillator tube and the stabilizer coil, and it still won't sync.—J.B., FPO, NY.

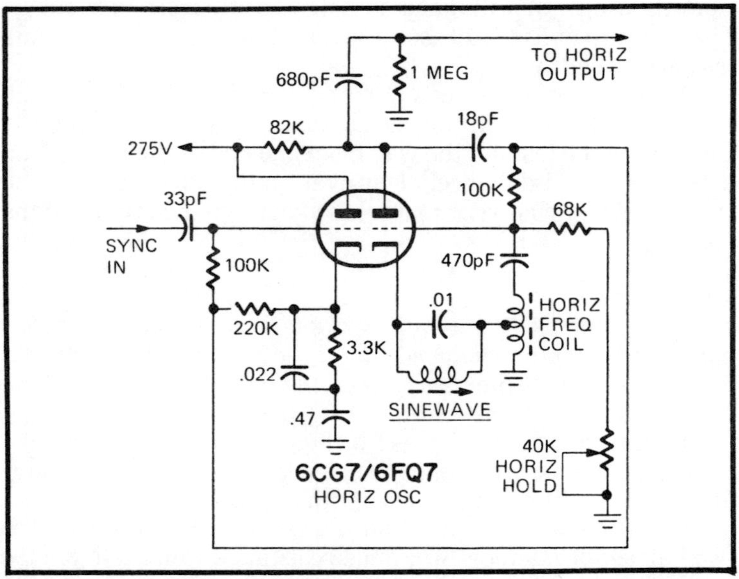

Fig. 2-20. This horizontal oscillator circuit is a fine performer when it's in good adjustment. But if you don't follow the factory procedure, you may wind up with multiple images.

This chassis uses a variation of RCA's famous *Synchroguide* circuit, and must be set up using the factory procedures or it won't work properly. Try this: 1. Connect a jumper across the terminals of the sine-wave coil (Fig. 2-20). 2. Ground the grid of the sync output tube; pin 9 of the 6EA8, on the same PC board with the oscillator. 3. Adjust the horizontal hold control until you can see only *one* picture. This will float from side to side, but if it will stand still for even a moment, fine. This means that the oscillator is able to free-wheel. 4. Take the jumper off the sine-wave coil. If the picture falls out of sync, adjust the core of the sine-wave coil until it locks in again. There's still no sync, remember, so the picture will float; get the sides of the picture straight, and it should hold fairly still. Shorting the sine-wave coil should cause only a small sidewise shift.

Final step; take the short off the sync-tube grid, and the picture should lock in very firmly. Change channels and see; it should snap in instantly if it's working properly.

Vertical White Line On Screen

I've often seen a thin horizontal white line on a TV screen, but now I have an RCA KCS-153 TV with a vertical white line! I thought the yoke had turned for a minute, but it hadn't.—O.W., Suffolk, VA.

You have just found one of the screwball symptoms that show up in transistor TV and puzzle us old tube-type technicians for a while. This is caused by an open yoke circuit! In tube circuits, an open yoke would kill the high voltage. However, in transistor circuits, the yoke is driven directly off the collector of the horizontal output transistor; there is no flyback winding (to match impedances) *between* the output transistor and the yoke. The high voltage winding is also driven by the same *pulse* from the output transistor that drives the yoke. So since you do have high voltage, the pulse is obviously present and normal! This clears the oscillator, output transistor, HV winding, HV rectifier, and so on, of suspicion. In other words, everything is okay except the *horizontal sweep*!

No Vertical Deflection

I just got hold of an old but good Sylvania Model 132 scope. It's been pretty severely fiddled with, but I've got most of it working. However, I can't get any vertical deflection. Vertical amplifier stages seem to be okay.—J.H., Houston, TX.

Check from the vertical deflection plate on the crt, which is pin 8, back through the wiring. Somewhere in there you'll find a switch marked VERTICAL DEFLECTION. It seems to be used to open the circuit from the vertical amplifier to the crt; for times when the signal must be fed directly to the vertical deflection plates.

It could be open or it could be defective. You can clean it, replace it, or if you don't need it, just jump it. This should give you vertical deflection. Sensitivity of this model, by the way, is 0.21V per inch.

Vertical Symptoms, Horizontal Trouble

The raster of this RCA CTC 55 slowly shrank, and just as it was almost gone, the circuit breaker tripped. It does this after about 10 minutes of operation. I found nothing bad in the vertical output stage; in fact, if I feed a 30V ac signal to the output grid while the raster's shrinking, I can get a full raster again. Only thing I can see is a gradual drop in the boost while it shrinks. This is weird.—A.J., Minneapolis, MN.

Right! Your vertical stage is okay. The raster-shrink is due to the drop in *boost*; this feeds the input-half plate, through the vertical size control. So you have a vertical symptom and a horizontal cause.

Find out why your boost is dropping. In quite a few of these chassis, this has been due to grid emission in the 31LZ6 horizontal output tube. It upsets the bias and makes the cathode current rise; this trips the breaker.

Horizontal Oscillator Too High

All I can get out of this Zenith 14M29 chassis is a high-pitched whine. Flyback and oscillator coil have been replaced. The horizontal oscillator is running, but at a very high frequency. What's going on here?—H.B., Riverdale, IL.

This set uses an electron-coupled horizontal oscillator, very much like the circuit used in many radios! Notice in the pictured circuit (Fig. 2-21) that the horizontal oscillator coil is tapped, and that the two halves are not equal. Also, the two capacitors connected across the coil are not of the same value.

Reversing the coil connections or the capacitors would change the frequency of the horizontal oscillator. Check this. At a guess, since the oscillator is running too fast, I'd say that you have the low-inductance (higher-frequency) section of the

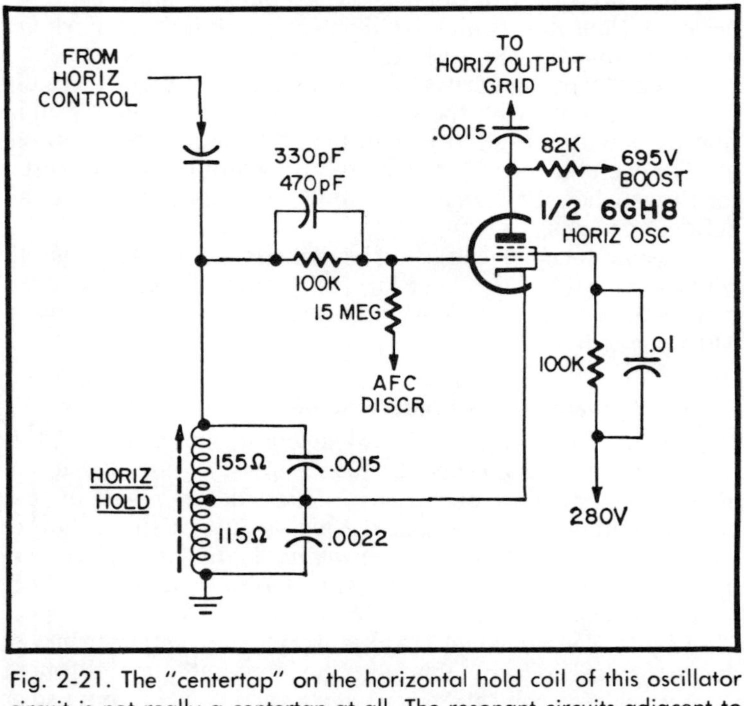

Fig. 2-21. The "centertap" on the horizontal hold coil of this oscillator circuit is not really a centertap at all. The resonant circuits adjacent to the tap are tuned to different frequencies.

coil "on top"; in the grid circuit. The diagram shows the correct connections.

Middle-Stretch In Raster

This is a new one on me. I've seen pictures stretch at top or bottom, but never seen one stretch in the middle. What causes this?—M.P., Del Rio, TX.

Most likely cause, the deflection yoke. Frankly, I don't know the exact nature of this defect, but I've cleared up quite a few cases of it by replacing the deflection yoke. Probably some odd short.

Not Enough Width

I replaced the deflection yoke in this GE M-760-CWD, and got the raster back. Works fine now but I don't have enough width. Need about an inch on each side. Width control doesn't help. I have a bad hum on all channels, too.—W.P., Carolina P.R.

Two possibilities here. One, low dc supply voltage due to a bad electrolytic in the voltage doubler. This could reduce the width and cause the hum all at the same time. Check dc voltages at power supply. Normal: -300V.

No. 2: If the width control doesn't have *any effect* when you move the core, it could easily be *shorted*.

No Horizontal Sweep

I had arcing all around the flyback in an Emerson 835A. I replaced it. Now I have a bright vertical line down the middle of the screen.—W.K., Hicksville, NY.

You have an *open* horizontal yoke. You are getting high voltage and vertical sweep, obviously. However, your horizontal yoke coil is open somewhere between the flyback and the yoke itself.

Pull the yoke plug and take an ohmmeter reading between pin 1 and pin 6. This will tell you whether the yoke windings are open; should be about 10 ohms. From either 6 or 3 you should read 4700 ohms to pin 1; this resistor is inside the yoke cover.

If these readings are okay, then recheck the flyback connections!

Chapter 3 Component Peculiarities

Transistors run cool—transistors have no filaments to get hot, so they last forever. These and other optimistic statements used to be found in all of the literature (including, I blush to admit, one book I wrote *before* I had a chance to do too much actual work on them!) If you find these two statements in a true-false quiz today, you'd better mark them *false*! I'm speaking mainly of power transistors now, but it applies to all types.

So, let's go over a few of the important things we need to know about this heat situation. Transistors do run hot, and if they run too hot, they don't last forever; their life can be measured in milliseconds! This will help when selecting and installing replacement transistors. The higher the power, the more careful we must be. Some definitions first:

1. Power dissipation. This has been a gray area ever since transistors first showed up. This term actually means the amount of electrical power lost—used up—in the **transistor itself**, running at full output. It is **not** the full-power rating of the whole circuit.

For example, a 50W transistor might be able to handle currents that would give a power output of 250W **into the load**. In this case, the total power consumption of the circuit would be 300W; 250W useful load power and 50W dissipated in the transistor. The diagram of Fig. 3-1 below shows this schematically.

Power lost in an electronic circuit usually shows up as heat. The transistor gets 50W hot, and this is the heat we must get rid of to keep it operating.

2. Maximum Power. Can a transistor with a maximum power dissipation of 50W actually dissipate 50W in service? The only answer to this is the classic "Yes, but..."

The maximum power-dissipation rating of a transistor is based on its maximum permissible junction temperature and

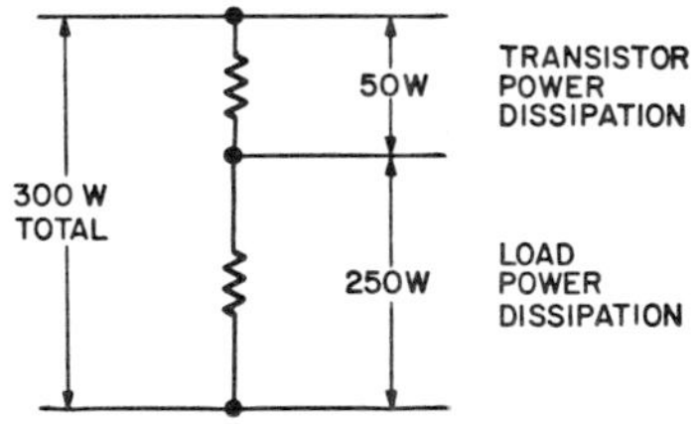

Fig. 3-1. Power dissipation in transistor.

its thermal resistance. The last term here means the ability of the transistor to transfer the heat from the junction to the case, and from case to heatsink. This is given in degrees Celsius (C). (To get degrees Fahrenheit from degrees C; multiply by 1.8, then add 32.)

Thermal Resistance

Let's pause briefly and kick "thermal resistance" around. We can figure it out, just as if it were an actual resistor dissipating a given number of watts. This has nothing to do with voltage or current, but we work in watts, which create the heat in the first place.

The closest analogy I can think of is a car's cooling system. The heat is generated in the engine block. From there it's transferred to the coolant (air or water, depending on whether you're driving a Porsche or a Ferrari). The coolant *transfers* the heat to the outside air through the radiator.

Naturally, there is a loss of efficiency in this process, and this is the thermal resistance of the system! If it were zero, the engine would always run at the same temperature as the outside air (which would be a little rough in Wisconsin in January!) So, a little loss is okay. Here, the junction is the "engine" (in the antique sense of "something that does something.") The mounting of the junction on the silicon chip is the "coolant" or transfer medium, and the transistor case plus heatsink is the radiator. Like the engine, we can't let the source get too hot. In the engine we melt down the bearings; in the transistor, we melt down the junction.

Figure 3-2 is a rough sketch of how this works. Now let's see how it works in terms of the only thing *we* can measure—the case temperature of the transistor. To get this, we take the maximum junction temperature (the maximum wattage we can dissipate), and multiply it by the thermal resistance, which depends on the case construction of this transistor. This figure is *subtracted* from the maximum junction temperature, and we come out with the maximum case temperature.

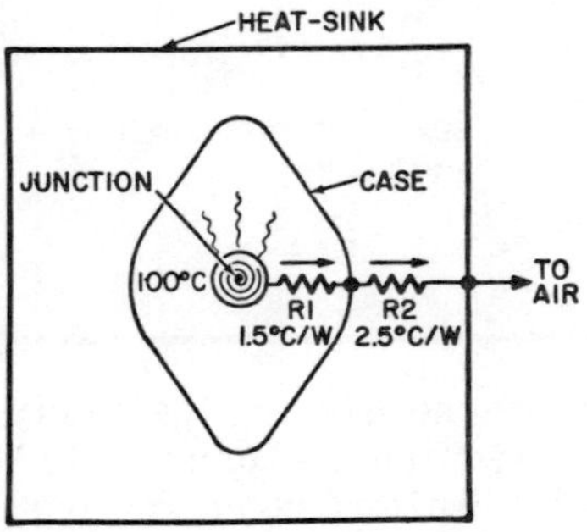

Fig. 3-2. Heat dissipation.

Our transistor has these characteristics:

Dissipation rating (max watts): 50W
Thermal rating: 1.5°C per watt.
Max junction temp: 100°C

Now, let's plug these in and see what we get: 50W flowing through a thermal resistance of 1.5°C/watt gives us 75°C slot right there. Subtracting this from the junction temperature, we come out with 25°C. So, this is the maximum *case temperature* we can use. It comes out as 77°F ($25 \times 1.8 + 32 = 77$)

Now, more problems. We put this transistor on a *heatsink* (radiator) which helps to keep it cooler. However, now we've got another figure to work in; the thermal resistance between the case and the heatsink, which isn't perfect either. A typical figure might be 2.5°C/W. This must be *added* to the transistor's internal thermal resistance and we come out with a big 4°C/W total.

So now we've lost a good deal of that nice fat 50W dissipation rating we started out with. In fact, the original 50W rating has fallen to only 10W. If only 10W flows through a thermal resistance of 4°C/W, we've lost 40 degrees. Subtracting this from the maximum junction temperature (100°C), we come out with a safe case temperature of 60°C, which is nice warm 140°F.

Much of our heat calculations must be based, of necessity, on the *ambient* temperature, which simply means the air temperature of the room where this gizmo is working. Normally, this will be about 70°F. If the air conditioner hasn't broken down or there isn't a brownout. However, outdoors in the summer, the ambient can easily exceed 90–100°F in the sunshine. So our heatsink has to be able to radiate enough heat to hold the transistor to the safe computed figure. Naturally, the higher the ambient, the lower the *difference* between heatsink and air, and the lower the efficiency.

So although we started out with a big transistor with a 50W dissipation rating, by the time we took off all of the deductions, here we sit with only a 10W rating! Remember that this is **not** the total power into the load; this is still the safe working rating of the transistor itself. This, of course, depends on the amount of power used by the load, the applied voltage and current, etc.

So when you're figuring out power ratings for transistors, original or replacement types, be sure to check the columns marked "Max Pwr Diss" and "Max Junction Temp," and the heatsink used. You'll be able to get the thermal resistance of the heatsink in most cases. If not, a ballpark figure of 2°C/W is pretty safe. Most of them will be close to that. For safety's sake, never overrate; underrate. Forced-air cooling and other means can help, of course.

CHECKING THE CHIP

One of my alleged friends once gave me one of those signs saying "An *expert* is one who knows more and more about less and less!" In some fields this is true, but the American electronics technician is just the opposite. He knows more and more about more and more! He has to. They keep throwing new things at us—color, transistors, and now integrated circuits. You name it, we fix it!

We are finding ICs in several things now and we'll find more of them as time rocks along. So let's take a look at some of the test methods and instruments needed to find the trouble and fix it. One of the first in common use is the integrated-circuit color demodulator, which they call "the chip." It has been in use for a couple of years and is doing very well.

Chip Checking

The chip used here is a "dual double-balanced synchronous detector" if you want the description. A typical unit contains 21 transistors and a lot of resistors, etc. However, we do not need to know this! All we need to know is the normal input and output signal amplitudes and wave shapes, and the dc voltages at various terminals, plus the normal current drawn. (We can't get inside the chip to do anything; all we can do is check it to find out if it is working.)

This is the basic problem we face—is the chip working or is the trouble in the external circuitry? So we must have tests

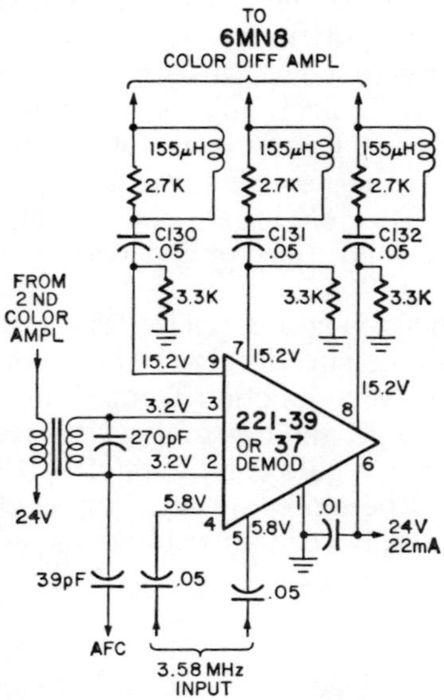

Fig. 3-3. Zenith color demodulator IC.

that will give us this data in the least possible time. Of course, when these ICs first came out, the regular number of perfectly good chips were replaced. However, it is not difficult to pin down the trouble with the right test equipment and methods.

The only reliable way of checking is probably the oldest in the business. It's simply "signal input/signal output" and dc voltages. We read the input signals, then the output signals. If they are normal, go on—the chip is okay. If these aren't normal, then we read the dc voltages at the IC terminals. If the dc supply voltage is normal, but the dc voltages at other points on the chip are not, then there is a definite chance that the IC is bad. All of the dc voltages found on other terminals (from the input supply) come "from within" the chip, and indicate trouble if they're off value.

The diagram of Fig. 3-3 is the circuitry of the chip in the 14Z8C50 series Zenith chassis. All of the normal dc voltages are shown. The color signals are fed into terminals 2 and 3 at a very low amplitude. You'll see a color-bar signal here, with a low-capacitance or crystal-detector probe on the scope, at about 0.75V p-p. The chip demodulates these signals, and comes out with the three color signals: red, blue, and green, at a good deal greater level.

The output signals will be the familiar "rocker" waveforms, just like those you see on the tube demodulators and color-difference amplifiers. They're fed to the control grids of a triple triode 6MN8 color-difference amplifier tube, at about 10V p-p. The green signal, as usual, will be somewhat lower than the red or blue. The color outputs from the chip are pins 7, 8, and 9. They can be easily identified by the external load resistors, 3300 ohms each, and the 0.05 μF coupling capacitors.

So there's the first test. Feed a color-bar signal to the antenna of the set and check for these signals with a scope and low-capacitance or crystal-demodulator probe. Actually, you can use a low-capacitance probe on the output. If you get the right signals here, there's no *need* to check the *input*.

Second-Stage Testing

Now suppose that we do not get a normal output. *Then* we check the input. If we see the normal, very low level signal here, but get no output, we read the dc voltages around the chip socket. To repeat, all of the dc voltages found here are developed inside the chip, except for the supply. On the specific chip in question, the three color output terminals, 7, 8, and 9, should read +15.2V. Terminals 2 and 3 should read +3.2V, 4 and 5 should read +5.8V. Pin 1 is ground.

If any of these voltages are off limits, for example, 7 and 8 normal, 9 low or high; check all external components connected to the faulty terminal. An open load resistor, leaky coupling capacitor, etc. could be the cause of this type of trouble. One good quick check for possible external troubles is simply pulling the chip. There should be zero dc voltage on all pins except the 24V supply, which should rise to about 26V or so.

Another good check for possible chip trouble is to read the current drawn. In this one, normal current is 22 mA ± about 1.0 mA. Current a good deal greater than this would indicate a short; quite a bit less than this, some kind of open circuit inside the chip.

Leaky coupling capacitors can be checked by lifting one end and rechecking the dc voltage on the chip socket. Consider the two 0.05 μF capacitors which feed in the I/Q CW signal from the transmission line to the oscillator. If either of these was shorted or leaking, the dc voltage would drop because of the low resistance to ground of the transmission line chokes, etc. Resistors can be checked with an ohmmeter.

You might think that these things run cool. They don't. The chip case will warm up in operation somewhat. If there is an internal short, it will probably get hot enough to make you withdraw the fingertip pretty fast. If the chip is "stone cold" after the set has been running for 5 minutes, it is quite apt to be open—or the 24V supply missing; check that first. This can be done from the top by taking a reading on pin 6. All leads of the 221-37 or 221-39 ICs are exposed, since it is mounted on a wafer base. Voltage readings and replacement can be made without pulling the chassis.

All in all, though these things are a novelty to most of us, I don't think they're going to cause us any more trouble than any other type of component. With a few logical tests and diagnoses, we can spot and clear up any kind of trouble originating in the ICs or in the external circuitry.

A DROPPING DIODE—WHAT IS IT?

What in the world is a dropping diode? Just what it says. A diode connected in series with the heater string of a hybrid TV set—the kind with four or five tubes in the high-power stages. You might say it's a heater rectifier, and in fact, it's marked FIL RECT in some schematics. True, in a sense. Actually, it looks to me as if it's used to drop the ac line voltage to about half normal. One case of this is in RCA's KCS-176A portable chassis.

The use of these can lead to some confusion, and some funny voltage readings, if we're expecting them. The heater circuit is in the diagram of Fig. 3-4. We're not rectifying ac to get dc, in the regular sense; note that there aren't any filter capacitors on the load side. What is happening here is that the rectifier is cutting off half of each sine wave. So, we get approximately "half-power." The *heating effect* of the current is reduced, referred to full-wave ac.

This gives us some unusual readings if we use our regular rms-reading ac voltmeters. Note the voltages shown along the heater string. The 33GY7, for example, shows 55V on one side

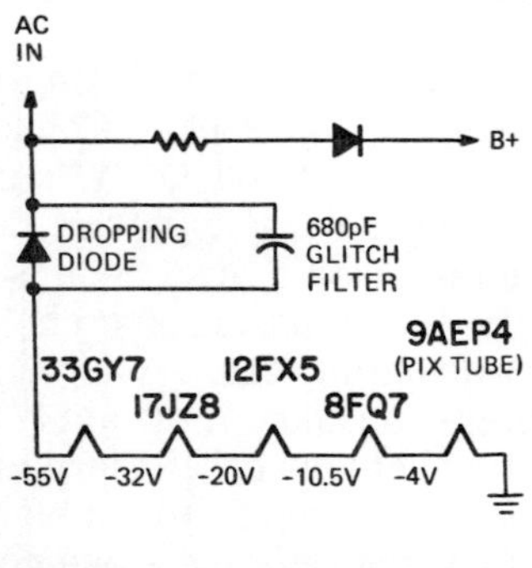

Fig. 3-4. The diode rectifies half of the input sine wave, reducing effective voltage to half.

and 32V on the other; total (55−32) = 23 volts! Yet, the tube works. Note that these voltages are read with a *dc voltmeter*; it shows −55V, etc. The normal ac drop across the heater string adds up to 76V, but we see only −55V at the input. The heating effect of the half-wave-rectified ac is enough to get the cathode up to the correct operating temperature.

We set up an experiment, using a 33GY7 tube, in a tube tester, on an adapter. Setting the heater at the rated voltage we got the readings shown in Table 10-1. We used a standard VOM, a FET VOM, and a digital VOM. On ac volts, the readings were correct. Then, one side of the heater was opened, and a standard silicon rectifier diode was connected in series. The heater voltage was not changed.

Now look at the readings! On ac volts, we get roughly half the applied voltage, which makes sense. Setting them on dc volts, we got about the same readings. (Oddly enough, the *emission* of the tube didn't seem to go down too much. Perhaps I didn't wait long enough for the cathode to cool down; tests took only about 3 minutes.)

This type of circuit can lead to some unexpected problems, if you don't know it's being used. In one, although the customer's complaint was poor focus, the tubes suddenly started glowing very brightly, then the fusible resistor in the B+ supply popped out. Checks showed that the diode in the heater string had shorted. This applied the full 120V line voltage to the heaters.

The diode was replaced, and the B+ checked. A heavy short was still there. This turned out to be the audio tube, a 12FX5, with a short. It's possible that the heater overload had caused this. Although the other tubes had been overloaded, too, they weren't damaged. Note: this diode seems to be used *only* in the "A" series of the KCS-176 chassis. In later runs, higher-voltage tubes were used, and a small dropping resistor (15 ohms) was added.

In other sets, you'll find other things used as droppers. In fact, anything having either resistance or reactance can be used. Small chokes can be used, but these are much more expensive than diodes; so are big resistors, nowadays. In some imported TV sets, a special nonpolarized *electrolytic capacitor* was the dropping "resistor!" Needless to say, this was a very odd value, and practically impossible to get. We finally began replacing these with stock resistors, which worked beautifully.

If you run into a set with some kind of unusual dropper, and you can't find a replacement, good old-fashioned resistors will solve the problem. You can find the value needed with Ohm's law. Don't forget to figure the wattage, too. Some of these will get pretty warm. Most can be mounted in some place on the chassis where they'll stay cool enough.

Suspect the use of some odd thing whenever you see that the heater voltages of the tubes do not add up to the line voltage. Here's how to get the right value; add up the heater voltages of all tubes in the string. Now, subtract this value from the line voltage at 120V. In the example above, this would be 120 − 76 = 44V drop, across the dropper. The current will be the same in all tubes, so you need only look up one type. Here, it's 450 mA. So, Mr. Ohm says that R = E/I, or 100 ohms. If you come out with an oddball resistance value, use the next *higher* stock value; this will give you a little "cushion" in case of line surges, and the tubes will never know the difference. You can run them at least 5% below rating, with full output.

The dropping resistor must have an ample wattage rating. Here, we'd have $P = I^2R$, or 0.45^2 times 100; comes out around 20 watts. Use a 25W type, allowing some margin as a safety factor. You can mount this on a couple of small metal brackets, close to the chassis for good heat transfer.

So, there you are. With good old fundamentals, and some scratch paper, it's not too hard to cope with some of the more unusual things you run into. Just keep your eyes (and your mind) open.

IC'S, TV, AND YOU

We're running into more and more sets using ICs lately. They have been "in and out" of quite a few makes, even the "unmodular" jobs, for the past couple of years. As usual, we take the things they throw at us, and develop test methods and equipment to cope with 'em. Let's look at a few do's and some pretty important don't's in this kind of work.

I'm not going to try to go into the internal construction of ICs. A schematic of the innards of an IC is pretty useless. So we go back to one of the oldest concepts in electronics—John F. Rider started it back in the early 1920s: Plain old signal tracing.

The IC is a sealed unit—a real *black box*. Some service manuals actually refer to the terminals as "connections to the *outside world*." So we have two ways of attacking the problem. Signal tracing, or "signal in vs signal out," and dc voltage readings. In my opinion, we should always use both of them; one to verify the data from the other. If we have normal signal in, and no signal or abnormal signal out, this is a good clue that there is something wrong in between.

Dc voltage tests are useful, if correctly interpreted. The majority of ICs seem to use one terminal as the power input. Internal connections distribute it to the jillions of parts inside. In all that I've run into so far, there will be several other terminals which show dc voltages. These come *from* the IC, through transistors and resistors, etc. These are a valuable clue.

So here we can get a small handle on the problem (the problem, in most cases: "Is the IC working?"). If we have the normal dc voltage at the power input terminal, and the normal dc voltages on terminals where it should be, this could mean that the IC itself is working. For verification, scope the signals at the signal-in and signal-out terminals.

For one example, in color demodulator IC, the normal input signal would be the reference oscillator signal, on at least two inputs, and the output would be the three color signals. Find out which terminals these are applied to, and where they come out. Check for proper peak-to-peak voltage and waveform. These are given on a lot of schematics, and should be given on all of them.

Let's say that we find normal inputs; plenty of 3.58 MHz oscillator signal, color signal from the bandpass amplifier okay. So we go and look for the three color output signals. The output waveforms of these, by the way, will still be the familiar "lazy-S" or rocker shapes that we've been seeing all along. If the set uses the old "video to cathodes, color to grids" circuit, they'll be exactly the same. In the "RGB" circuit, where color and video are fed to the cathodes, the patterns will be inverted, but still of the same shape. When making these tests, always use a color-bar signal on the input so that you'll have a fixed, recognizable pattern.

If the dc voltages are normal but signals are abnormal, check any external parts that are connected to that terminal. For example, in one circuit the three color signals come out through a resistor and capacitor. An open resistor or shorted capacitor could upset the signals, and not have too great an effect on dc voltages. If you find two normal color outputs, and one is zero, and the external parts are okay, this would be a definite indication of IC trouble.

In one typical set, the supply voltage is +12V. On the three color-signal output terminals, dc voltages of about 5.6V are developed inside the IC, on each of the terminals. If two of these are normal, and the other zero, or 12V, this means internal trouble. Be *sure* to check first for anything *external* which could cause this: a solder bridge, shorted capacitor, and so on.

TEST EQUIPMENT AND METHODS

I believe that the minimum test equipment you'll need for this kind of work will be a bar/dot generator, a good scope, and a VOM with a very high input impedance: either a FET VOM or VTVM. The voltages won't be very high. So, the voltmeter should have a couple of good low-voltage ranges; say 0 to 2 and 0 to 10 volts, capable of being read to within tenths of a volt. A wideband triggered-sweep scope makes things very handy. However, older scopes can do the work, if properly operated.

One of the most important things in this kind of work is the test prods and probes on your equipment. You simply cannot use the old "phone-jack" test prods on the closely spaced pins of an IC with safety. Pin spacing of a 14- or 16-pin DIP (dual-inline package) IC is only a few *thousandths* of an inch, and the old tips cover two or three at once. If you short two adjacent pins, the results are apt to be disastrous.

For maximum safety, use needle-point test prods. I've noticed that a lot of test equipment makers are providing very sharp pointed prods on the late models, equivalent to the needle-point prods. These, of course, have been around a long time. You can change your present equipment for needle points, with no trouble. Somewhere, I have seen a slip-on adapter for the old prods, with a needle point.

Your testing methods are going to have to change, too. No more happy jabbing in the general area of where you want, and hoping you hit it. When you take voltage readings on an IC socket, steady your hand on the chassis, and bring that tip down very carefully on the terminal.

If you can rest your hand on the chassis, this helps. If they have thoughtfully provided a couple of hot terminals right where you want to go, put a small piece of foam plastic over them. Get a piece about an inch thick. This should be enough to insulate you, and also help to keep you from knocking over heatsink tabs on transistors, and other stuff that's near the area where you want to work.

One final thing. We're used to counting tube-socket pin numbers clockwise, from the bottom of the chassis. The round-can ICs still count in this way, but you seldom work from the bottom. The DIP ICs are numbered *counterclockwise, from the top*. Start with the IC set so that the locator notch on the case is to your left, with the long side of the case parallel to you. Now, go *down* and start with the left pin. This is No. 1. Count along the bottom row, then turn the corner, go up, and keep counting until you get back to where you began. This is easier to show than describe (Fig. 3-5).

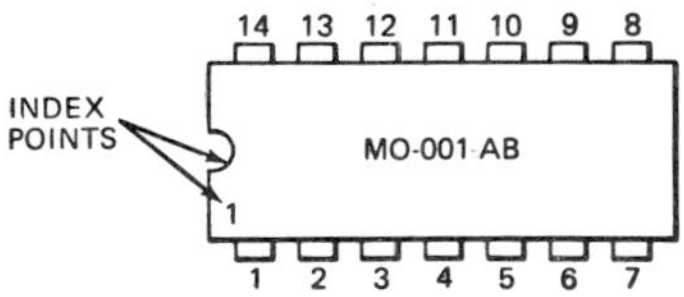

Fig. 3-5. Top view of a dual-inline IC. In some cases pin 1 is marked by a dot or the maker's symbol.

As I said in the beginning, these are all "new and different" to most of us. However, I still have my abiding faith in the American service technician. We've taken everything they have thrown at us, in stride, without missing a step, and we'll do the same thing with these little monsters.

REPLACEMENT PARTS

You hear the factory men say "Always use exact duplicate replacement parts, our part number _______". In a few cases, you can find suitable substitute replacements. In others, brothers, you'd *better* use the factory replacement part. If you don't, you can "build in" a real stinker of a problem. I ran into a couple of very good examples not too awfully long ago, to illustrate this point.

The subject was a Magnavox T933, which showed only blue and green. It had been worked on just before I got it; complaint: "wrong colors." So I ran a gray scale on it and, after forgetting that in these sets you must set the screens with the brightness control turned *off*, got a fairly decent monochrome picture. No color at all.

Up bar/dot generator and at 'em. Now I've got color all over the place. Take generator off, hook up antenna, fine color picture. What happened? Oh, oh. Intermittent. Turn off, cool, turn on—no color now. Up scope. Touch plate of 3.58 MHz oscillator tube with probe. Pow! Color again, and oscillator signal. Repeat test, with same results. Turn off, cool, no color, apparently due to loss of oscillator signal. Touch *anything* in the circuit, and the oscillator starts.

Now sit there and think of a really clever way of checking this without touching it. Okay; place the scope probe near (but not touching) the plate. Recheck. No oscillator. Wait. No signal, after several minutes. The oscillator will *not* start by itself. Touching the plate, even with the tip of a pencil, starts it, and it's right on the money.

Try running the regular oscillator setup tests. These produce very odd results, and the oscillator still won't start. Maybe this is the crystal. Oscillators refuse to start because of insufficient feedback, and in this one, the crystal is the main element in the feedback loop.

Having a couple of crystals on the bench, I try one. Same problem. Try another crystal; Same, only worse this time. Find brand-new crystal, for Motorola replacement. Try this. Hmm. Won't work either. Okay, try the acid test; go down the street to the Magnavox dealer, and get the right crystal.

Put this in, turn it on, and presto, it works! Run the setup adjustments, and everything is copacetic. For a recheck, I tried the Motorola crystal once more, and one of the others. Same thing; they won't work. So I make a resolution. When I see what I suspect will be crystal trouble, from now on I will always use the *right* one! There are enough differences in circuitry, constants, so that the wrong crystal will refuse to work, or won't work as it should!

That got the color problem. However, while working on that, I'd noticed some odd beats on the screen. I ignored these, since they didn't seem to be associated with the color problem, and weren't. Now I went after them. The tuner was apparently oscillating; low-band stations were okay, but on the high bands the thing was as hot as a pistol. Touch the lead-in at any point, on high-band stations, and the picture went all to pieces.

This is ordinarily due to improper neutralization, in the "Neutrode" tuners. I'd seen it in certain RCA tuners, and cleared it up by readjusting the neutralization. However, this one didn't *have* a neutralization adjustment. Let's try a new rf tube. Pull the tube, which is a brand-new one. Hmm. Tube layout calls for 6HQ5, and this one is a 6HA5.

On a hunch, I got a 6HQ5 and tried it. Perfect. No oscillation, no squiggles, nothing. Put the 6HA5 back, here we go again. Apparently, this substitute tube had just a little bit *too much* gain, making the tuner unstable. Installing the correct tube type fixed the whole problem.

The moral of this is that if you must substitute for any part, check the set's *performance* after the replacement is in. *This* is the final criterion. I *have* substituted one tube type for another, with good results. I have substituted a lot of transistors for the original types, also with good results. I have also tried substitute transistors and tubes, and found out that they wouldn't work satisfactorily.

A Zenith 14N22 had a severe case of vertical jitter. Full, linear raster, all control reactions normal, plenty of sync, etc. but it jittered. (I did not suspect the tube, because of full raster, etc.) The scope showed an odd pattern on the output waveform of the vertical integrator (see Fig. 3-6). There was a tiny glitch on the scan portion of the waveform. This was intermittent.

After checking all of the parts I could think of, and replacing the sync separator and integrator, I went to the Zenith dealer and got a 6FM7 tube (as usual, I didn't have the one I wanted). Try this. No jitter! No glitch on the waveform. Put the original tube back. Jitter again. Get still another 6FM7 tube; no jitter.

This one, I am *not* going to try to explain, for I have no idea of the cause. However, that one particular 6FM7 tube *would* cause the jitter, and the others wouldn't. There are "fixes" given for this, including adding a diode in the oscillator cathode, but these had no effect. The jittering tube checked perfect, of course.

So when you run into mysterious symptoms like these, try a new tube, or better still, a couple of new tubes, and see if this won't help. Even though the original seems to check out perfectly, and works almost normally, try some more tubes. Sometimes, it pays!

SOLID STATE "TUBES"

One of the first things they said about semiconductor devices was "They will replace vacuum tubes!" They haven't quite made it yet, but they're working on it. In fact, the only place where you can make a one-for-one direct substitution is in the rectifier socket Let's look at a few of these.

Plug-in direct replacements for the old standard full-wave rectifiers have been available for quite a while. They started with the 5U4, etc. Now, you can get them to replace practically every rectifier tube, from the old faithful 80 up to 866s and 872As; transmitting-type rectifiers capable of carrying current up to 1.25 and 1.75 amperes, at working voltages up in the thousands.

When making this type of substitution you may have to add a small series resistance. It compensates for the lower voltage drop across the solid-state rectifier. Without it, your dc voltage may run high enough to exceed the working voltage rating of the filter capacitors.

The next one that came along was a solid-state substitute for the focus rectifier tube, in color TV sets. The type R-2AV2

Fig. 3-6. Vertical integrator anomaly.

solid-state focus rectifier is a plug-in replacement for the original 1V2 and 2AV2 rectifiers in the receiver.

Up until now, it has been common practice to make high-voltage rectifiers by stacking individual solid-state diode "chips" or discs in series, to get the higher peak inverse rating needed to withstand high surage voltages. Variations in distributed capacitance, voltage breakdown, and switching-time between individual diodes have caused uneven distribution of the voltage drops, and eventual breakdown. This was quite a problem in the first focus rectifiers. However, since this is a dry circuit (practically no current at all), it wasn't too hard to overcome.

The high-voltage rectifier in color TV posed a problem, for the reasons just given. The very high peak voltage, up to 35,000 volts, and the fast rise time of this pulse (from nothing up to 35 kV in about 7 μsec) made it difficult to keep everything lined up! About 1968, Motorola came up with a successful HV rectifier. They called it the "SSHVR" (solid-state HV rectifier). They wanted it to get rid of the last chassis tube in the Quasar series. This is the only one of the whole group that isn't really a plug-in replacement for tubes like 3A3, etc.

To get the best performance out of this rectifier, a new flyback transformer should be used. It must have more parasitic stray capacitance to match the lower capacitance of the SSHVR. The socket isn't compatible; the SSHVR has only a plate-cap type connector on each end, which is all it needs. Improvements in device *matching*, together with a great improvement in corona-proof packaging and encapsulation, made this unit possible.

The EDI R3A3, R-3AT2 and R-3DB3 Solid-Tubes are direct plug-in replacements for vacuum-tube rectifiers and are ideal in cases where the rectifier filament winding may be defective.

Chronologically, the next tube to be replaced was the damper. Electronic Devices, Inc. has just announced another direct plug-in substitute. The R-DW4 Solid Tube is designed as a direct replacement for 6DW4, 6CK3, 6CL3 and 6BA3 tubes in color TV circuitry. Because of the very high pulse voltages in this circuit, it was pretty difficult to work out a practical replacement for the vacuum tube here, but with improvements in technology and fabrication techniques, they made it.

Introduced in 1971 was the solid-state voltage multiplier unit. RCA, Magnavox, Sears, Sylvania, Zenith and others began using them. They're made in different configurations. RCA uses a voltage quadrupler, and most of the others use voltage triplers. The basic circuit action is similar to the well known voltage doubler, which has been around for a long time. The first time I ran into it was in an old DuMont, back in the 1950s! (I still remember opening up the HV cage and wondering what the heck they were doing with two 1B3s!)

Voltage-multiplication in the HV stage has several advantages. For one, the flyback transformer does not need that great big HV winding! Not nearly so big as before, anyhow. For example, to get 24 kV with a voltage quadrupler, the flyback pulse can be only 6 or 7 kV instead of 30–35 kV. This reduces, if not eliminates entirely, any chance of X-radiation from the HV rectifier, since you have to get up above 20 kV before you find any perceptible X-radiation.

There is another nice little fringe benefit. On most of the multiplier assemblies, you can tap off the focus voltage, after the first stage. There goes the separate focus rectifier. In addition, the focus voltage will automatically remain right where it must be—a fixed percentage of the applied HV. About 20% is average.

For typical service work, you can always replace a tube-type B+ rectifier such as a 5U4 with a pair of silicon diodes. Just pull the old tube, and tack two of the little 1.0 ampere 1000V diodes to the plate and cathode terminals of the socket. (It is highly advisable to connect these right. Otherwise, it can be embarrassing. If you're using some of the bead type silicon diodes, get out the magnifying glass, and look for the band marking the cathode end. Otherwise, pop goes the fuse.)

Afterward, be sure to check the surge voltage. Connect the dc voltmeter to the input filter capacitor, turn the power on, and read the surge voltage—it should be greater than before. If the surge goes above the safe rating of the filter capacitors, you'd better add a small surge resistor, either between the rectifier output and the input filter, or between the ac supply and the rectifier anode.

In some cases, the extra 20–40 volts of B+ will be a big help. In some of the older sets, this may even help you to get that last ¼ inch of width that you need!

GETTING TO KNOW VARACTOR DIODES

Many manufacturers are using no-switch TV tuning these days. They're doing it with voltage-variable capacitors, or *varactors*.

Let's see what a varactor is and how it works. As simply as I can put it, it's a plain diode. This, like all diodes, has a certain capacitance between cathode and anode. When we change the reverse bias on it, the capacitance changes. That's all there is to it. The drawing shows a diagram of the thing; a little crude, but this is how it works.

We have a block of semiconductor material. Each half is doped with impurities, one P and the other N. The interface between the two is the junction. Now if we reverse-bias the thing, it stops being a "diode" (something that conducts current one way but not the other) and becomes a completely open circuit for dc. (Capacitor is an open circuit; remember?) It will have a capacitance between the two elements. The value of this capacitor depends on the size of the block. (Incidentally, these are just like zener diodes in one way; they are always operated with reverse bias. However, they do not conduct dc in normal operation, as a zener does. If they do, they're shot.)

Okay, now we have capacitance. We can connect it across a coil, and it will tune it—make it resonant at a given frequency. We need to be able to vary the capacitance so that we can tune the circuit where we want it. In a TV tuner, it doesn't take much.

Now how can we tune it? Simply by applying more or less reverse bias to it. This will cause the "charge layers" in the diode to repel each other, and move farther apart or get closer. Nothing actually moves, of course, since this is a solid-state device. However, the charge layers do actually move farther apart, or nearer. This is exactly the same as moving the plates of a variable capacitor closer or farther apart. The capacitance changes.

Now we've got something very useful! We can hook this gizmo across the coil of a TV tuner and vary its resonant

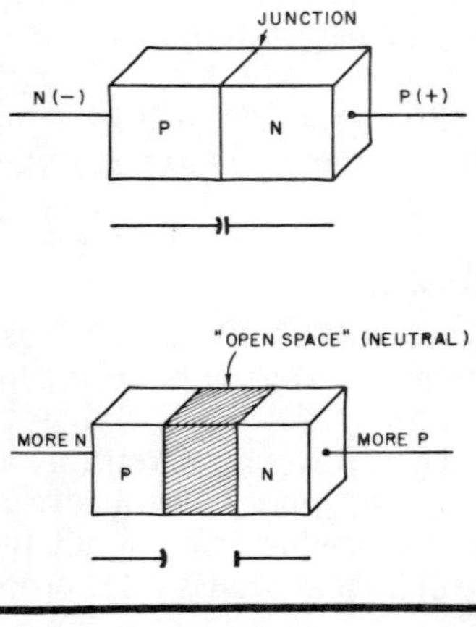

Fig. 3-7. Increasing the reverse bias on a diode widens the depletion region, and the effect is the same as increasing the dielectric strength or moving the plates apart on a real-life capacitor.

frequency without all of those frightfully complicated multicontact switches we are all so familiar with! All we need is a source of variable dc voltage; battery and variable resistor, in effect.

The guy in the back row with the red necktie, as usual, brings up a point; "How about the presence of dc in an ac (tuned) circuit?" Makes absolutely no difference! Proof? Look at the two-winding i-f transformer of any tube radio; +200 volts dc in one winding, −3 to −4 volts in the grid winding, and it works. By properly *isolating* (bypassing) the dc, it will have no effect on the tuning action of the varactor.

So we can make a very simple TV tuner. The channel selector consists of a set of precision resistors, mounted on something like the familiar "turret" selector switch, so that only one is in the circuit at any time. Only two contacts instead of 8 or 10!

To get a 4-circuit TV tuner (antenna, rf, mixer, and oscillator), we simply connect four varactors, one across each of the tuning coils, and feed them all from a common dc supply. For example (and this is close to the actual voltages used in the UHF section of one current tuner), if we switched +22V onto the varactors, the tuner would "stop" on ch. 65, or thereabouts. (Here's a point to remember in all varactors: the *greater* the reverse-bias voltage, the lower the capacitance; the smaller the reverse bias, the greater the capacitance—the closer the charge layers. Example: 0.5V, minimum capacitance).

For those who are wondering; yes there are *trimmer* adjustments on the coils of all of these tuners, so we can still "tune" them up, for the proper bandpass, tracking, etc. In fact, in all of the varactor tuners I've seen so far, the varactor is in parallel with a small capacitor; sometimes fixed, sometimes adjustable.

So basically, these things are jsut like the tuning devices we've been used to. The big thing is that they're a lot simpler and a lot easier to work on. In fact, I wish that all present tuners could be magically converted to varactors! (Especially that intermittent dog in my own set!) Before we leave that point, "No, Irving, you can not (practically!) *convert* an existing tuner to varactor operation! Not without a complete redesign of the tuner.

The diagram of Fig. 3-8 shows a simplified basic circuit of one stage of a TV tuner using this type of switching between high and low VHF channels. (This happens to be the rf input, but all of the rest will look the same.) Remember, in this we will run into the combination of "ac and dc effects," but don't

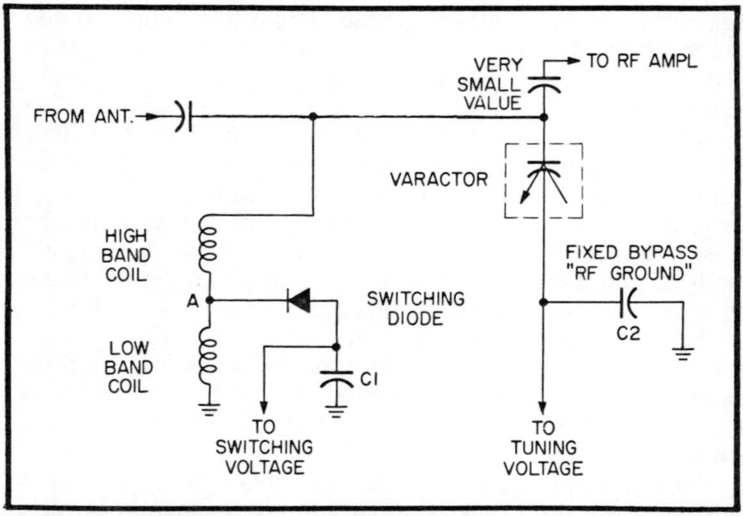

Fig. 3-8. There are many symbols for a varactor; the one shown in this circuit is one of the least common. Most show a recognizable diode and a recognizable capacitor rolled up into a single symbol.

let 'em confuse you. They actually have no effect on each other.

The coil is the same as that used in those all-too-familiar switch tuners (incremental-inductance type). For tuning channel 2, the whole coil is in-circuit. Here, this would be both "low" and "high" coils. To get to a higher-frequency channel, we simply short out parts of the coil with a selector switch. On channel 13, all we have left is one miserable little half-turn of wire.

In the circuit pictured, we have a single varactor diode, which tunes all of the high channels or low channels, with a single coil for each. It does this by varying the "tuning voltage" on the varactor. The high-band coil is the top one in the diagram. The "low band coil" is both of them. We will call the lower part the *low-band coil.*

Now, note the diode connected to the middle of the coils. This is it; the thing that switches the tuner from low to high bands on VHF. Let's see how. For low-band operation we want both coils in the circuit. So we feed a high reverse-bias voltage (the "switching voltage") to the diode. This makes it a completely open circuit. And C1 is not in the circuit at all (Remember C1: this is the key part!)

To cover the low channels, 2 to 6, we vary the varactor dc voltage. This causes it to change capacitance, and away we go.

To change to the high channels we have to get that high-inductance (comparatively speaking!) low-band coil out of there. As you know, the high VHF channels work at such high frequencies that the "inductors" wind up (that wasn't intended as a joke, and it wasn't) as only a half-turn of wire apiece. We can't have those great big 4-turn low-band coils cluttering up the place.

We get rid of them, just as in the older tuners, by shorting them out. Only now, we short them all at once. How? By connecting a good big *bypass capacitor* from the junction of the two coils to ground! That's why I told you not to forget C1! It is actually a feedthrough capacitor, but it is a 0.001 μF unit that is a very dead short to VHF. So if we "close a switch" at point A, we have made this point a very good rf ground, and now we have nothing but the high-band coils left in the circuit. The varactor can "tune" these coils to any desired high channel, by varying the dc tuning voltage.

How can we "close the switch?" By applying a high forward bias to the *switching diode*. When we do this, it conducts heavily, and becomes a nice fat short circuit, switching the capacitance of C1 from the center of the two coils to ground.

If anyone's interested, we can use the same varactor for tuning both high and low bands. We simply feed it more or less bias voltage. Remember, a typical varactor can be changed from about 2.5 pF to 12 pF simply by changing the dc tuning voltage from 0.5V to +28V. With properly designed coils, this much change in capacitance is plenty.

There you have it. Simple, huh? If we remember the basic principle of operation we can get along very nicely with the latest model tuners. In fact, they ought to be a lot *easier* to work with than the older models! For instance: suppose a tuner works fine on the low band, but not at all on the high band. The switching diode is open, or the switching voltage is not there. The whole coil is in-circuit at all times. If it were just the reverse—high band fine, low band nothing, the switching diode is probably shorted.

In other words, it'll be just as it is now—if we know how the thing works, we'll have a lot less trouble finding out why it doesn't work.

REPLACING HARD-TO-FIND TRANSISTORS

The invention of the transistor about 25 years ago started something. For one thing, it opened the floodgates for a simply tremendous stream of literature written about it. The mathematicians had a field day! They happily poured out a flood of formulas and then developed more, and more, and more—the h-parameters, the y-parameters, in addition to the few *simple* formulas that we need to determine the performance of the thing in a given circuit.

Besides this, they came up with the incomprehensible concept of "holes." This may be valid, but it served mainly to bring about a state of total and utter confusion in the minds of those of us who must work with the things. I can attest to this, not only from my own personal experience, but from thousands of letters in the Service Clinic mailbag.

Unfortunately, this kind of math is based on the premise that these devices are precision instruments! In other words, they're all alike. Later thinking contradicts this.

To be practical (to work as predicted), one would suppose that transistors would have to be capable of precision reproduction. In plain English, this means that every device of a certain number would have precisely the same characteristics; gain, leakage, current rating, and on and on. This is simply untrue. A group of 6CG7s all have identical characteristics; a group of 2Nxxxx's do not.

You *can* get a group of transistors with almost identical characteristics. You can *not* build them that way, though. You make them, and then you *select* those with almost identical characteristics. All you have to do to verify this statement is to read the spec sheet on any transistor. You'll find that they show a *minimum*, *maximum*, and *typical* value for every parameter of the device. For *beta*, as an example, you may find it varying from 50 to 150. This is called "parameter spread," as well as other names, some not nearly so polite. Even these must be read at a fixed temperature, 25°C. All parameters vary wildly with temperature; leakage, beta, breakdown voltage, and on and on.

In a large transistor factory I visited some time ago, the final-test machine was a monster almost 100 feet long. The transistors were plugged into little adapters, and traveled down a track on the top. At each stop, they were tested. The first few were for shorts, opens, etc. Later, they were checked for all other parameters. I asked the friendly soul who was giving me the guided tour, "What's that last station, the one on the very end?" He grinned and said "When they get there, we raise the price three dollars and mark 'em MIL-SPEC!"

The transistors that were kicked out into bins in the middle of the machine were those that weren't precisely on spec, but within a given percentage *variation*. For example, if the bogey value of beta was 100, the machine could be set to kick out all that fell within a variation of +50 to −50 (except for those that read *precisely* 100; these went on to the last stop).

You can verify this, too. Just buy 100 commercial transistors and check them for beta. One of my friends has shown me this. Out of 100 "identical" transistors, 50 will be right on the typical, 25 above, and 25 below.

The wide tolerance of these things is amazing. When we were in radio school, we worked things out to four decimal places (which I always thought a bit unnecessary). When we got out into actual service work, we found tolerances of 10% and even 20%, and felt that this was being a little slack. Normal tolerances in transistors are really something; one authority gives the figure of +100%, −50% for the beta alone.

From the preceding gloomy statements, you may begin to wonder how they could *ever* build anything with such inconstant devices. (So do they, at times.) However, there is a way. Later thinking has worked out means of compensating for the variations. To make a practical transistor circuit (one that works), the design engineer simply *designs the transistor out of the circuit.* He does this by choosing his circuit constants so that the variation in parameters is simply *swamped out.* If this transistor has a normal leakage current of 500 μA, he sets his normal resting current at, say, 2.0 mA. This is so much greater than 50 μA that the leakage-current variation won't cause too much change.

The next step in this process is the liberal use of feedback, both current and voltage. By doing this, he can swamp out the effects of any beta variation, and hold the overall gain of the circuit constant under worst-case conditions.

The final step is "derating." This is simple; we've been doing it for years; we called it "safety factor." Example: if a circuit designer needs a transistor with a maximum collector voltage breakdown of 50V, he uses one with a breakdown rating of 100V or even 200V (if *production control* will let him). By derating voltage and current ratings, the junction will run as cool as possible. This eliminates the worst single cause of parameter variation; a rise in the *junction temperature.* This makes every parameter of the transistor vary tremendously.

Right about now, you are probably saying, "What's this all got to do with *me*? I'm not designing the stuff; I just fix it." That's the point, brethren. To make *practical* repairs on this stuff, we have to know how it works, and above all, the working *limits* of the devices. If you understand these, you can make substitutions and in most cases come up with a circuit that works better than it did before it failed. All you need is a copy of any one of the transistor replacement guides, which give the ratings for the whole line of general replacement transistors. I have six of these on my desk now, and there are several others. They're the most consulted reference books I have. All of the data you need is there.

You know that one of our worst problems is locating replacement transistors. We never open up a piece of equipment that we don't see a whole new set of numbers, if the things have any numbers at all on 'em. The makers of general replacement transistors claim that only 50–60 of their transistors will replace thousands of OEM numbers. This is quite true.

There simply isn't all that much *difference* in transistors. For example, if you need an i-f or tuner transistor in a TV set, one general replacement transistor, such as RCA's SK-3018 or Motorola's HEP-728, can be used to replace any one of literally thousands of OEM types. There are several others, of course, in other lines; I just picked those at random.

All common transistors could be lumped into a very few classes, with a very small set of the really important parameters. The *least* important to us is the beta. (More on this later.) The most important, in my mind, is the collector breakdown voltage and maximum collector current rating.

A properly designed transistor stage will be amply derated, as I said before. My pet authority on the subject recommends that a transistor should never be used at a collector current that is more than 40% of its maximum current rating. The same thing would apply to the collector voltage breakdown, though I like to double this, at least, and have used a factor of four.

The most controversial statement that I've made so far is probably the one about the beta being the least important parameter. Like everything else in our business, this isn't unqualified. I did say that in a *properly designed* circuit, it wasn't too critical! This is due to the fact that the beta of all transistors is far from being a constant thing. It will vary wildly with temperature, current, applied voltage, leakage, and you name it. In the properly designed circuit, these variations will be swamped out so that wide variations of voltage, current and even junction temperature will have only a small effect on the overall *gain* of the stage.

Besides these things, we have the wide parameter spread; often as high as 75% ± the bogey value. So, even if you are using an exact factory replacement, it's possible to get one with a beta that differs quite a bit from the original. (You may get one that's precisely the same, but then you may not. I've checked a few "matched pairs," and found them highly unmatched!)

In actual service work, and in a long series of experiments using general replacement transistors, I've found that this will work. I have substituted as many as five different "replacement" types into TV sets, and found no trouble at all. If you can find a number on the thing, and it's listed in the replacement guide, the chances are it will work very well. The only time I've ever run into trouble is with the occasional typographical error.

There is one case where you may run into some problems. This is in the multistage direct-coupled audio amplifiers. If you replace an output transistor, and the resting current is incorrect, try replacing the driver transistor. The original driver may not have precisely the correct beta. (Although, as said before, if the designer swamped the circuit out, this won't upset it. Try using types with either a higher or lower beta, and see. It may be the way the stage was designed in the first place! To check, read the resting current of the output transistors.

You can make your own experiments. The next time you get a TV set with plug-in transistors, try a few from your general replacement stock. Try picking one that *looks* as if it would work. The chances are that it will.

READER QUESTIONS

IC Lifter

What's the easiest way, if any, of getting ICs out of their sockets? I broke one today when those wee little pins hit my thumb.—M.R., St. Petersburg, FL.

The best gadget I know of is the little one that I got with a Heathkit. Just a piece of heavy-gage sheet metal, a bit less than ¼ inch wide. Just enough to let it slip between the pins of a DIP IC. Bend one end to a right angle, about ½ inch long. The other end can be any length you want, as in the sketch (Fig. 3-9).

To use it, slip the short end under the IC and carefully pry one end up. Then pry the other end up by moving the handle the other way. You can "rock" it up out of the socket a fraction of an inch at a time, without bending any pins.

The *Techni-Tool* people make a plier-like device with jaws, which is also very handy.

Burnt Cathode Resistor

I'm just getting started in electronics. I found a little radio, with the cathode resistor of the audio output tube burned

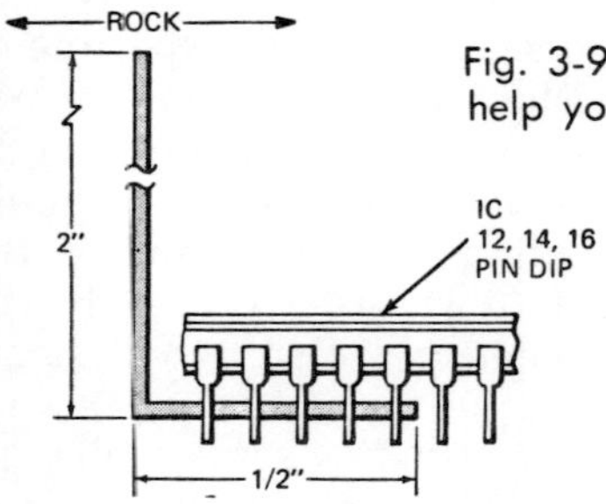

Fig. 3-9. Simple IC lifter. Other types help you get ICs back in their sockets.

up. The tube was good. Why did this resistor burn?—K.P., Beaver Dams, NY.

Because there was too much current drawn through it. That sounds a little elementary, but it is true. So if the tube is good, there is only one thing that could cause it to draw so much current. Its gird must be very highly *positive*. Most likely cause of this, a shorted coupling capacitor to the grid from the plate of the preceding stage.

Remember this reaction; you can use it in every kind of electronics work, from little radios all the way up to color TV.

Transistor Collector Connections

I've got a silly one! I replaced a bad output transistor on a little import set, and now it won't work. I can't read any collector voltage on the tranistor case; got plenty on the supply lead! No smoke, just no sound.—R.Q., Waco, TX.

Oh, ho, ho! You have just been zonked by the same thing I was not too long ago! This is a bolt-on transistor, right? And, the original had a lot of insulating goop under the transistor and all over the screws, right? Well, when they say "insulating," they aren't kidding!

Take out the mounting screw that makes the collector connection. Clean off all of that insulating goop, and then put a small outside-star lockwasher under the screw head. Tighten it, and then take a resistance reading between the supply lead and the transistor case. I did, and with the first hookup, I had the biggest fattest *open circuit* you ever saw! Screw was tight, too.

Replacement Transistors

I need a pair of replacement transistors for a Craig Pioneer Series 8 tape player. In fact, I need two matched pairs. Originals are 2SB473. Found one listing in a replacement guide, but the cases aren't the same, so they won't work.—G.P., Hepzibah, GA.

The 2SB473 transistors are in the TO-66 case, which looks like a small TO-3. Sylvania's ECG-131-MP is a matched pair that will replace these. You can also mount the flat-case HEP-243, etc. transistors in the TO-66 mounts, if you can locate one in a germanium type, which the originals are.

Substitute For 12R5

The vertical output tube, a 12R5, went out in this GE 14P1215 portable TV. Not having one, I replaced it with a 12CA5. Worked for about 5 minutes, then a white band showed on the bottom of the screen. Not too long after this the 12CA5 went out. All adjustments react normally, but the tube won't hold up.—A.E., Neptune City, NJ.

This looks like one of those cases of "same basing but different characteristics" in the two tubes. The 12CA5 has a maximum power dissipation of only 1.5 watt; the 12R5, 4.5 watts. Normal cathode current in this set is about 40 mA, which is above the rating of the 12CA5. I'd advise using an exact replacement tube.

If you still get the foldover, check the 12R5 coupling capacitor. This causes most of this type of trouble.

Circuit Breaker Pops

The circuit breaker opens up on this 12-inch RCA HSK-T1, which I built from a kit. I've checked everything they told me to, with no results. The only way I can get it to hold is pull the high-voltage rectifier. If I hold a neon lamp near the plate lead, it glows brightly. I've tried a new high-voltage rectifier; no help.—M.R., Gregory, MI.

Most likely cause is some kind of short or leakage in the high-voltage lead from the rectifier socket to the picture tube. Since you eliminated the chance of a shorted high-voltage rectifier by replacement, this is about all that's left.

Loss of Volume Control

A Westinghouse BP19A770 TV has an odd volume control problem. Most of the time, the volume control won't reduce the volume at all. Full on. You can turn it off and on, and then it works for a while.—A.S., Northfield, NJ.

This problem is almost certainly caused by a poor *ground* connection on the control itself. All volume controls like this are "audio voltage dividers." If the ground opens, all you have is series resistance, and this isn't enough to hold down the volume.

The control itself is on the front panel, connected to the chassis through a shielded cable. The shield of this is used as the ground return for the audio signal. This does not go to chassis ground in this model, but to the top of a 330-ohm resistor in the 6AQ5 tube's cathode. Trace this, and fix the bad joint.

Replacement Transistors

I need a set of output transistors and a driver for a Wards GHJ-949B stereo. The outputs are marked "SPS-0122" and "SPS-0121." The driver is "036-9800." There's some confusion in the parts list. The outputs are the flat-pack types, but the replacements listed are TO-5. Do you have any data on these?—K.J., St. Paul, MN.

RCA SK-3083 and SK-3054 would work for the output pair, and an SK-3025 would probably make a good driver. Cases match the originals. When you put in the new outputs, be sure to check the emitter resistors, bias diodes, and anything else that could have been damaged when the originals went out.

Video Detector Blows

I'm having trouble holding video detector diodes in a Motorola 551. They blow out about once a month. We've checked the i-f transformer and the whole circuit, and can't find any leakage which could cause this.—E.C., Milton, WV.

We went around once before with this problem, in a certain model of an old set. The answer in that case was the same as this; use a detector diode with a higher reverse voltage rating. We replaced the tiny glass diodes with a good old 1N34 or 1N70, which have a rating of 60 pV or more. These held up.

I have run into the same problem in some sets made for different companies by Sharp. Same cure. In these, the diode is in the last i-f transformer can; a tight fit, but possible.

Horizontal Stabilizer Coil Won't Work

This GE M110YBG has a peculiar problem. Horizontal sync won't hold. The horizontal oscillator coil doesn't have any effect at all, no matter where I turn the slug. Any suggestions?—L.A., Hobe Sound, FL.

Yes sir! One of them would be a careful check of that horizontal stabilizer coil. This very definitely *should* have some effect on the frequency. The coil may be shorted, or the 0.0022 μF capacitor may be open. One other good possibility: Take the core all the way out of that coil and check it. I have found a few of these with half the core broken off and missing. This throws the coil so far off resonance that it has no effect.

Short Life Transistors

The output transistor in this Mayfair FT-1024B tape recorder keeps burning up. The original 2SB178s had been replaced, and the replacements also burnt up. I replaced these, and now the new ones fry if I leave it more than about 60 seconds.

The base voltage is too high; −0.8V instead of the listed −0.1V. Collector voltage starts to drop as soon as it's turned on, and keeps dropping. Checked resistors, etc; no luck. I'm baffled.—T.L., Berkeley, CA.

I think your high base bias is the key to the whole thing! This is far too much forward bias, and as the transistor is driven into conduction, things will get worse. Note that this stage is clamped by a voltage-divider network from the −9V source (see Fig. 3-10.). This is composed of a 4K and a 150-ohm resistor with a small thermistor shunted across it. Base goes to the junction through the secondary winding of the transformer.

If the ground leg of this voltage divider *increased* in value, this would make the base voltage go up, (which it is doing). You have replaced both of the resistors, so this leaves us only one thing to suspect—that thermistor. I believe you'll find that this is either open (unlikely) or not properly soldered (very likely)! Taking off part of this *parallel* circuit makes the net resistance go up, which would give you just the kind of reaction you have. I have found quite a few small amplifiers like this, with thermistors open due to poor solder joints.

IC I-F With Problems

This Pioneer SX-1500T stereo receiver has a couple of problems. No multiplex, and the receiver cuts out when the "mute" switch is turned on. I can't find a schematic. Any ideas?—W.G., Madison, AL.

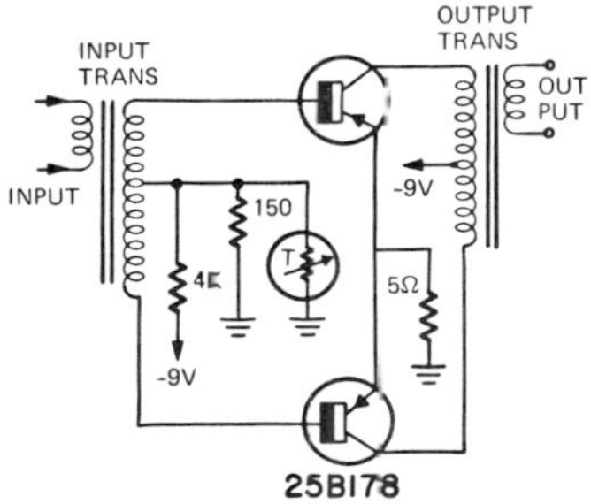

Fig. 3-10. Thermistor-biased output circuit.

No original ones. So I wrote to the company, and a nice man promptly sent me the service data, plus a suggestion that this was probably a bad μA703 IC. They use four of them in the i-f strip. He suggested other sources for the μA703s. Sylvania makes a replacement (ECG-703A) and there are others. (Look in the ads in the back pages! You'll find them there, too.)

Replacement For 21HJ5

It's very hard to get some of the odd tubes you find in older black & white sets! Right now, I'm hunting for a replacement for a 21HJ5, used as horizontal output in a 16M3 Admiral.—R.J., Michigan City, IN.

This is a Compactron; you might check with the nearest GE distributor to see if they have this one listed. If not, check through the GE tube manual. I see quite a few 21V, 600 mA types in this category. Most of them would require rewiring of the socket, although the tube characteristics look very much the same. (See Fig. 3-11.)

A 21HD5 tube looks pretty good. Same mutual conductance, current ratings, etc. Only the cathode would need rewiring; it's connected to different pins. Just watch out for *internal* connections and jumpers, which could cause shorts.

Parallel Diodes For Heavy Load

I have an RCA CTC 40, with a shorted horizontal trace diode. Ordered replacement, but got part No. 131475. This is rated at 6.5 amps. Original was a No. 126857. The high-voltage came back, all voltages checked normal, picture fine. The diode got very hot.

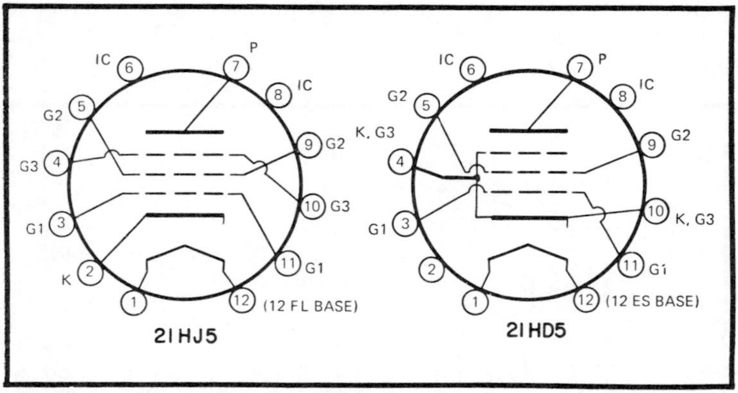

Fig. 3-11. The 21HD5 is a good replacement for the 21HJ5, but some socket rewiring is necessary.

Checking current, I read about 10 amps through it! Rechecked everything else. Okay. Ordered another diode, and got 131475 again. So, just for luck, I hooked both of them in parallel, crossed my fingers, and sent the set back.

It's been more than a week now, and I haven't heard anything. What would you have done?—E.S., Corydon, IN.

The same thing!

Remote-Control Transmitter Seen On Scope

I can hold a hand-held TV remote-control transmitter near the vertical input of my scope and see the signal. No electrical contact; I just aim it at the probe. Within ½ inch, the trace fills the screen. This is an ultrasonic signal, isn't it?—M.H., Los Angeles, CA.

Yes. About 40 kHz. However, many of these transmitters use a condenser microphone transducer. These things need a high dc voltage to work. So, there is a pretty strong *electric field* around the output. This will easily couple to a sensitive scope input.

Some manufacturers give you this as a test for transmitter operation. Some of these units are actually driven at ½ of the receiver frequency; the microphone diaphragms are designed to act as frequency doublers; they resonate at 40 kHz, but are fed 20 kHz by the circuitry.

Breaker Pops With Good Diode

Here's a weird one! If I remove diode D1 in this Admiral 5H10 chassis (Fig. 3-12), the set works; good high voltage and focus. If I put D1 back in, the breaker trips. Even if I take off the loads, and unhook the degaussing coil, it still does it. What is this?—M.H., Del Rio, TX.

Check that thermal switch. I think you'll find it is grounded, or stuck closed.

Hot (New) Flyback

This Admiral L33L31 came in with a burnt flyback. I replaced it with a new (bleep) exact duplicate. First replacement heated up and went out in two days. Got another one like it, and it's overheating. Cathode current of 6DQ6 is below normal. The only thing I can see is the resistance of the high-voltage winding; the spec sheet calls for 550 ohms; I read 620 ohms.—C.G., Alamogordo, NM.

While this is pretty rare, I am beginning to believe that your new (bleep) replacement flyback is defective. That high resistance reading on the high-voltage secondary could mean that it was wound with too small wire. I'd recommend

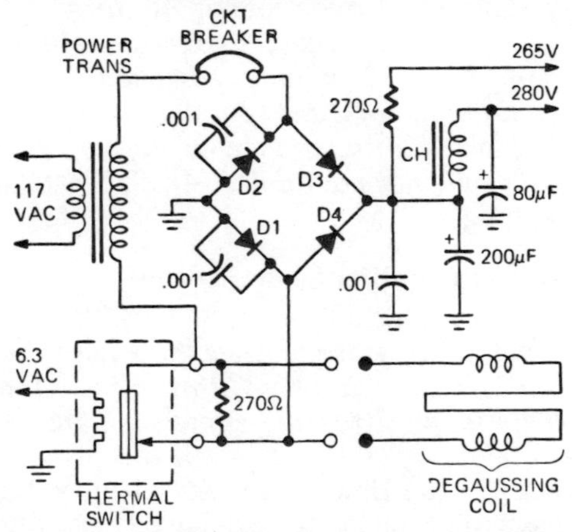

Fig. 3-12. Admiral 5H10 chassis power supply with circuit breaker.

replacing it with an Admiral part. Also, write to the (bleep) factory and let them know about this. Without field feedback, they can't tell.

Bias Diode Reversed

I cleared up some other problems in this Westinghouse V-2486-2 chassis, and I've got one left. This is a bad foldover from the bottom, in the vertical sweep. The bias on the output stage isn't right. Should be −15V and the best I can get is about −5V. I've checked the parts in the bias network, and I get +15V on the electrolytic capacitor.—G.M., Greensboro, NC.

You shouldn't. This voltage should be 15V *negative*; see diagram (Fig. 3-13). All of the other parts seem to be of the right value, but recheck that bias rectifier *diode*; it must be reversed. Some of these little diodes are very hard to identify, but if you're getting a positive voltage out of it, it has to be in backward.

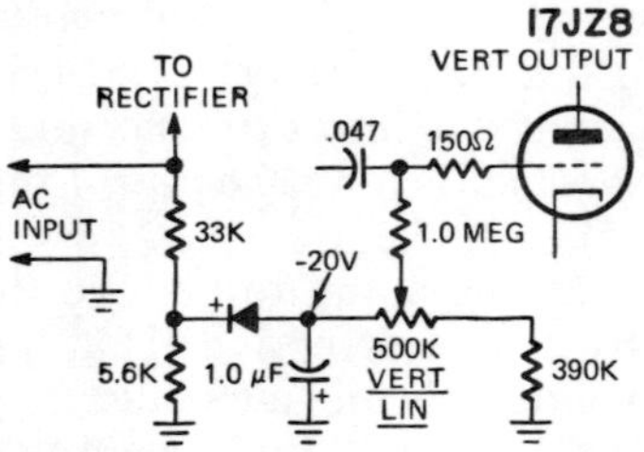

Fig. 3-13. Diode reversal caused foldover.

This would cut down the negative bias on the grid of the 17JZ8 output section and cause the foldover.

Resistor Burnout

The 1500-ohm resistor in the first i-f cathode circuit, in a Zenith 15L33 chassis, burns up. I've changed the tubes and checked all resistors; no luck.—B.T., Delta, IA.

This sounds a little basic, but the resistor is carrying too much current! Since the only thing that causes current flow in this resistor is the *tube*, something must be wrong there.

Not the tube itself, since it's been replaced. So, this leaves us only one thing—the grid bias. If this is "too far positive," the tube *will* draw excessive plate current. These little frame-grid tubes can handle surprisingly high currents, with incorrect bias.

The bias for this tube comes from the agc. Check the cathode voltage, **and** the control grid voltage. Note the high "offset" cathode voltage (see Fig. 3-14). The cathode is +42V from ground, but the grid is +35V. These are no-signal voltages. So the net bias on this tube is −7V.

If something is wrong in the agc which lets the grid go *more positive*, it will cause the tube to draw a heavy plate current and burn up the cathode resistor. (In case anyone is wondering about why the 22-ohm resistor doesn't burn up, too, it's easy. It's so little that it doesn't develop the I^2R drop (power) across it that the 1500-ohm one does!)

The Misplaced Point

The picture won't stay in horizontal sync on this Motorola TS-597. The hold control is supersensitive; pix falls out just the second you move it. I tried new capacitors across the horizontal oscillator coil/hold control. If I use the 0.22 μF on

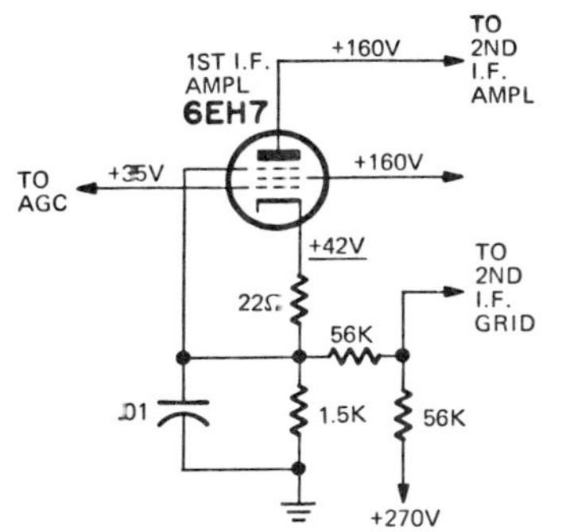

Fig. 3-14. Ground-leg cathode resistor repeatedly burns up; problem could be defect in agc.

the lower half, as shown in the schematic I have, it won't oscillate at all!—C.C., Johns Island, SC.

Two things here. The 0.22 μF capacitor across the coil should be a 0.022 μF (see Fig. 3-15). Someone has moved a decimal point. Incidentally, when replacing these capacitors, be very sure that you get the right one across each section of this coil. You can get some strange and wonderful effects by reversing them.

Two: your *oscillator* is working. This is an afc problem, if it won't hold. Check the dual diode unit, and the 150 pF sync coupling capacitor.

Flyback Resistance Different

I've just replaced the flyback in an RCA CTC 17X. Works very well, but my cathode current on the 6JE6 is running 230 mA, and I can't get it below that. Horizontal efficiency coil dipped, high-voltage regulator current about 1.4 mA, boost okay, and so on.

I used a well known replacement flyback. I noted that the resistance of the high-voltage winding is given as 660 ohms, and this one is only 500 ohms. Would you suspect a short, or something like that?—H.B., Hayward, CA.

Under the circumstances, no. I can't see a short in the high-voltage winding with only 230 mA of cathode current in the 6JE6. This is too close to being in the ballpark. I'd say that the lower resistance of this winding could be due to the use of slightly *larger* wire; if your high-voltage is up to normal, then the *inductance* almost has to be correct.

For the excessive current, check the "high-low" ac line switch. It might be in the wrong position.

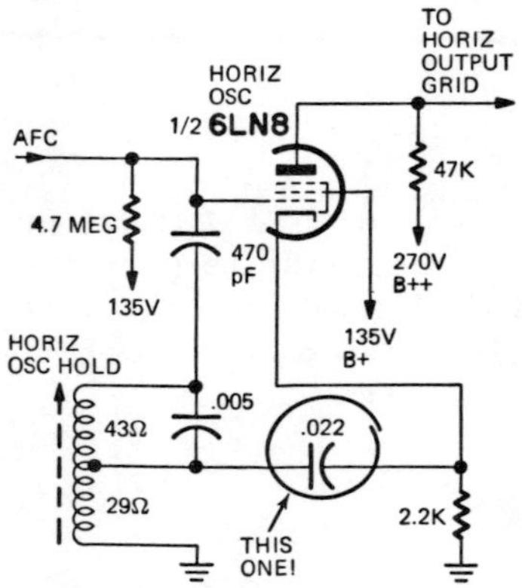

Fig. 3-15. Incorrect capacitor values in an oscillator circuit can keep the oscillator from working or pull it off frequency.

Flyback Replacement

If you have to replace the flyback in a Zenith 19DC20 hybrid chassis, you can save a great deal of time. The new flyback will come with the wires already soldered on it. *Don't* use them! Unsolder the original wires at the flyback terminal board. Take the old one out, put the new one in, and then remove the wires one at a time, and resolder the original wires. They'll be of the same colors.

If you do it in this way, you won't even have to pull the chassis. Otherwise, you'll have to pull the chassis, and spend quite a little time threading some of those wires underneath and getting at some tight places.

Hot Output Transistors

I've replaced the output transistors in a Zenith stereo, 20AT30Z. Now the new ones get hot and the tone is lousy. Driver transistor is okay. I changed it for the one in the other channel. New transistors match.—B.D., Billings, MT.

You have bias trouble, I'd say. Check the bias diode, marked CR451 on the board. If it is *open*, you'll lose the proper bias; the output stage will be unbalanced and get hot. If this is it, you can replace it with a plain silicon diode, such as RCA SK-3030. Your midpoint voltage (at the speaker connection, before going through the capacitor) will settle down somewhere around 14V. It must be half the supply voltage, which is about 28V.

Video Detector Blows

I've got a Truetone MIC3919C portable, which has blown three video detector diodes in about two months. I can't find any trouble in the set; no shorts in the last i-f transformer, etc.—R.Q., AZ.

Try replacing the original glass diode with a 1N64. This is how we cured the same trouble a long time ago in an old GE model with the same symptoms. Or use an RCA SK-3091 that has a 40V peak rating. Some of the imported diodes seem to have this problem. They short, without apparent reason. So use a higher-rated diode.

This one is in the last i-f transformer can. The old ceramic 1N64 will be "push-fit" in here but it can be done.

RCA Trap Won't Work For TVI Reduction

I've got a problem with interference (TVI) between stations. We're close to a strong local channel 10 station, and when we try to set an antenna for four other stations about 30 miles away, the channel 10 signal interferes! When we try to

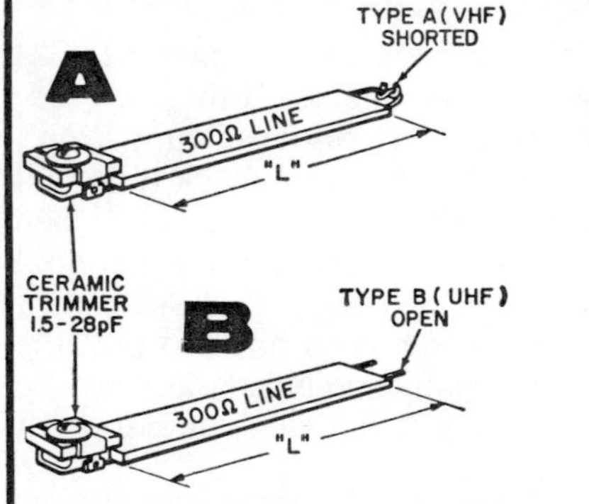

Fig. 3-16. Tunable TV suckout traps.

get channel 8, we always have channel 10 signal floating around in the background.

I've tried RCA's little trap, shown in their service publications, but it doesn't work at all. Any ideas on what we might do?—A.P., Tarpon Springs, FL.

Yep. Your're probably using the original RCA trap circuit, with a 4⅜ inch length of 300-ohm twinlead, shorted at one end, and a 2.5–14 pF trimmer across the other. If you'll look closely at the diagram (Fig. 3-16A), you'll see that it "tunes sharply from 40 to 170 MHz."

Channel 10 is 192–198 MHz! So the original trap is out of its range. Try the later versions (see Fig. 3-16B) from the April 1967 *Plain Talk.* Cut your channel-10 trap to only 3 inches of twinlead, and I think you'll find that it works a lot better. Also, use a 1.5 to 28 pF trimmer (low-loss ceramic type). Tape this to the lead-in, preferably between the set antenna terminals and the tuner, as close to the balun as possible.

Replacement Pix Tube

This Olympic 3P48 imported TV seems to have a bad pix-tube heater. It goes on and off at odd intervals. Voltage supply, socket, and so on check out okay. I can't find a replacement listed anywhere for it, and the company doesn't answer my letters.

This is a "310GB4" tube. Do you know of a substitute?—C.C., Cornwall, Ontario.

After much searching of characteristics lists, tube manuals, and so on, I may have come up with one (Fig. 3-17). The characteristics of the original are: 12-inch rectangular, 110° deflection angle, 12V heater at 75 mA. The heater in this little set is fed directly from the +12V supply for the transistors.

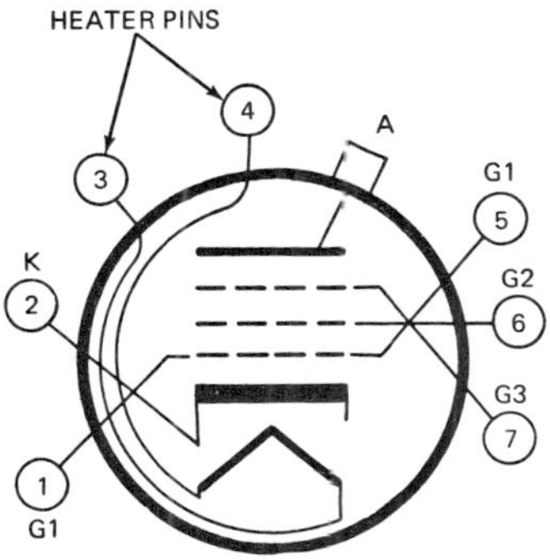

Fig. 3-17. Sylvania's 12BVP4 has 7GR base; tube is substitute for 310GB4.

Wonder of wonders, I located a tube in the Sylvania technical manual (13th Edition). It is a 12BVP4, and all of the characteristics match, even to the 12V 75 mA heater. Also, and even more unusual, the base seems to be the same. It is a "7GR" base, 7 pins. So with crossed fingers, I'd say that it ought to work.

Safety Precautions Cracked Pix Tube

I have an odd problem. This 21-inch color tube has arced, apparently between the 2nd anode and the degaussing coil shield. It cracked the bell of the tube! (The customer says he fell asleep while watching it, so he doesn't know how long the arcing went on.) Anyhow, now the thing has a crack all around the 2nd anode button, 6 inches in diameter, but it seems to still have vacuum!

I'm afraid to take it out, for fear that it will implode and wreck the chassis. How can I release the vacuum to play safe? Tube has no dud value.—J.M., Orlando, FL.

You're right; play it safe! Try this; take the back off the set, and lay it face-down on a cushion. Now, get an old quilt, and pack it inside the cabinet, all around the bell of the tube. Leave the neck sticking out, but wrap the quilt up around it pretty well, tying with string if necessary. Now, carefully take the pix tube socket off.

If this is one of the big-base round tubes, carefully drill a 1/8 inch hole right through the center of the locator key on the base (Fig. 3-18). This will hit the exhaust tip of the bulb, and you'll hear a loud hiss when it breaks. This will let air in slowly and the tube will then be safe.

If it's one of the new rectangulars, with the small base, you can crush the locator pin with a pair of gas pliers. This is very

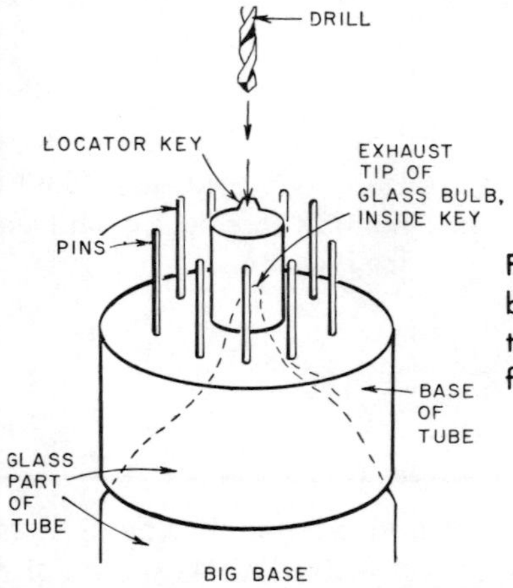

Fig. 3-18. Relieve vacuum in big-base tube by drilling through keyway. (Wrap tube first!)

thin plastic. The chances are that you'll break the exhaust tip when you do this. If not, you'll be able to see it, and you can snap it off with the pliers. Figure 3-19 shows how it looks. You might not have much vacuum left, but it never hurts to play it safe.

Replacement Switch

The on-off switch on this RCA KCS-160 is damaged. I'd like to replace it with one of the push-pull types. Don't find a listing for a replacement.—J.F., Media, PA.

This is a special dual switch (see Fig. 3-20). S1 is the regular on-off switch in the ac input. However, S2 is a completely *separate* switch. It is connected to the ungrounded

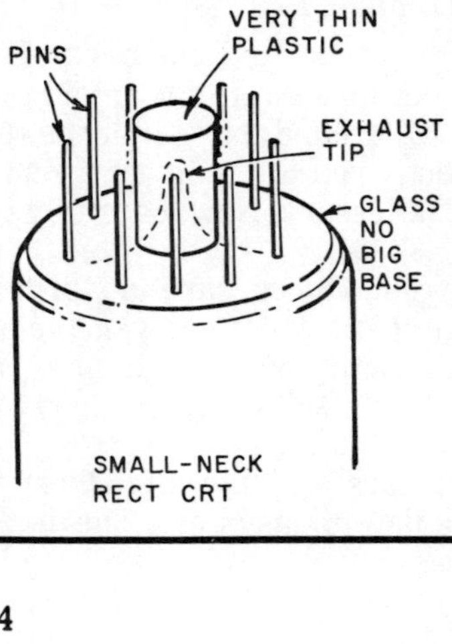

Fig. 3-19. Crush the thin plastic locator key with gas pliers (rectangular tubes). Be sure to cover tube with quilt.

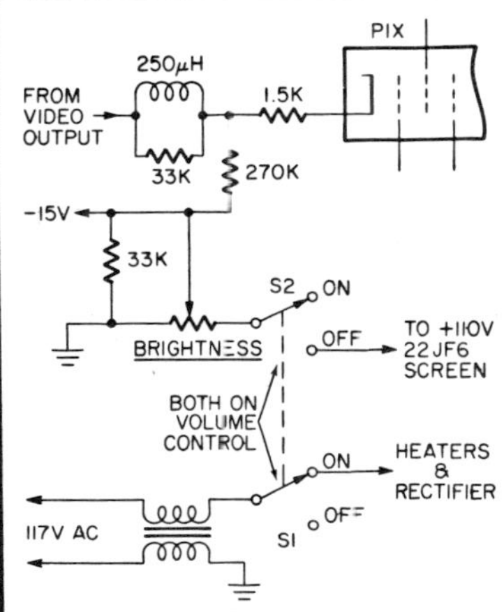

Fig. 3-20. Ganged power switch, RCA KCS-160.

end of the brightness control and to a +110V dc source at control and the picture tube cathode, thus discharging the HV quickly and killing the spot.

As far as I know, no one makes an exact duplicate. RCA part number is 119191. In an emergency, you can use any stock 1-megohm volume control with an SPST switch and leave the spot-killer circuit tube, it doesn't do much good anyhow!

Burnt "Plate Load"

The plate resistor of the pincushion corrector tube of this Setchell-Carlson U802 burns up; so does the 15K pot that feeds this resistor. I changed the 6GK6 pincushion-corrector tube, but that didn't help. The raster goes out of focus and pulls in.—J.A., Rome, GA.

May I make a small correction? The 15K pot and 1500-ohm resistor *do not* feed dc to the pincushion-corrector tube. The dc plate voltage is supplied through the primary winding of the pincushion-phase *transformer*, to the right of the tube in the diagram (Fig. 3-21). If you'll notice, the dc path to the resistors is *blocked* by the 0.0047 μF capacitor. Nothing but pulses *should* come through here.

I'd suspect that 0.0047 μF capacitor of being shorted.

Indicator Lamp Failure

The neon lamp indicator in an Allied (Knight) KG-711 depth finder kept burning out. The original is a type NE-2. They last only a few hours. The output transistor, a 2SB463, went out on me, and I replaced it with a 2N301. The circuit (Fig. 3-22) looks fairly simple.

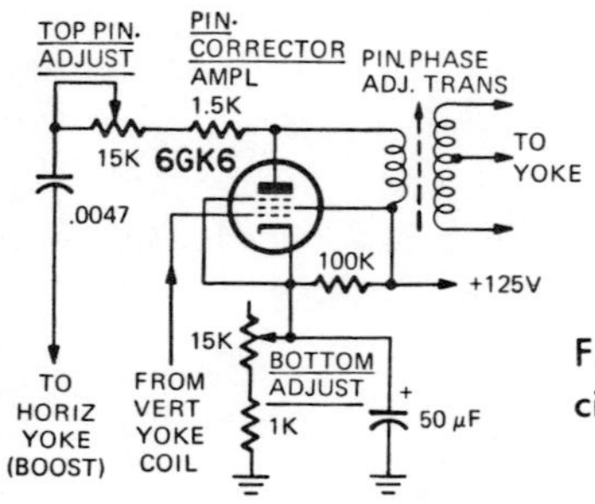

Fig. 3-21. Pincushion correction circuitry, Setchell-Carlson U802.

I finally got mad and replaced the NE-2s with NE-2Hs.—K.W., Sheboygan, WI.

Right. I'll take both fixes.

Focus Problem

After I replaced the picture tube on this Catalina 122-770B, it worked fine. I was converging it, and all of a sudden I lost focus. Smoke started coming out of the high-voltage cage. The flyback was pretty warm. I tried new tubes; same thing. 6JE6 cathode current went off-scale on 250 mA range.

Nothing helped; horizontal efficiency coil, focus adjustment, had no effect. The raster is full, but badly smeared; no focus at all. Do you think the flyback has gone?—A.S., Chicago, IL.

Not yet, if you still have a full raster. Don't leave it on too long, or it probably will be. This sounds very much as if your focus transformer is shorted. As you can see from the diagram (Fig. 3-23), it is shunted right across terminals 1 and 2 on the flyback.

To check, just disconnect the focus transformer from the flyback and try again. If your 6JE6 current goes back to

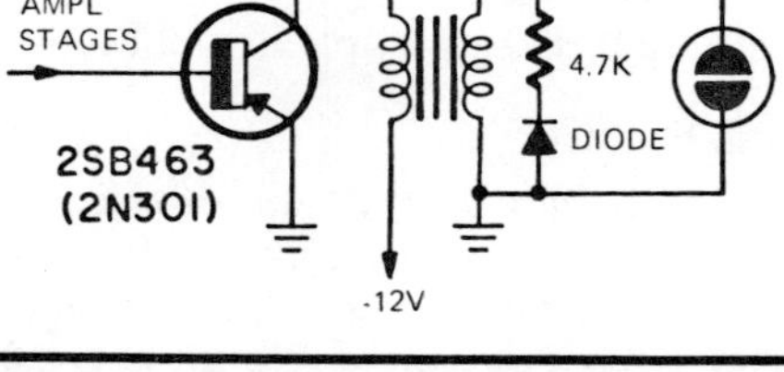

Fig. 3-22. NE-2 burnout problem solved in this depth finder by substituting NE-2H.

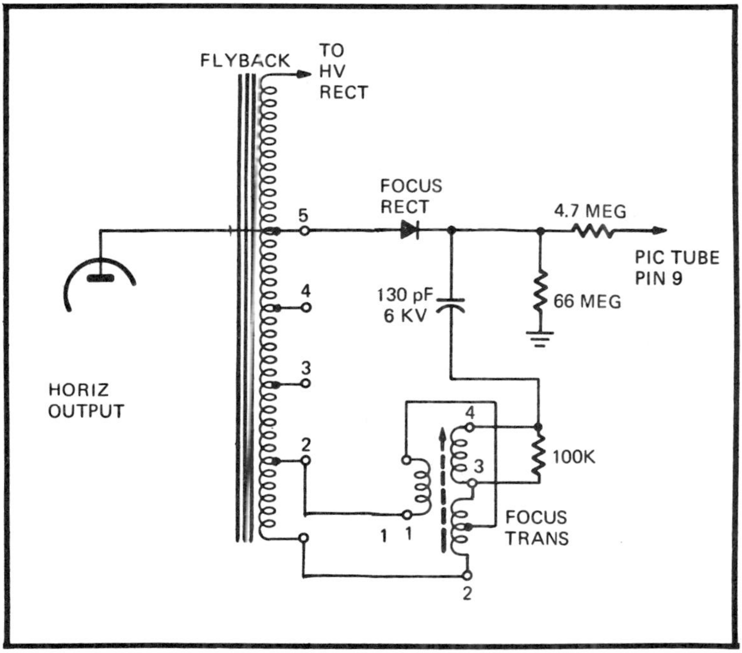

Fig. 3-23. Focus transformer in this set is connected across pins 1 and 2 of flyback. A short in the focus transformer will affect the flyback transformer.

normal, and the flyback stays fairly cool, this is it. You'll probably see much better focus if the transformer was shorted; this is only a *control* on the focus voltage.

I've got no focus voltage at all on a Zenith 25LC30. The focus rectifier socket has burnt up. This is riveted to the flyback cage. Can I replace this with a solid-state focus rectifier?—R.M., Mena, AR.

You sure can. An RCA SK-3066 can be soldered in place inside the flyback cage, with little trouble. Connect the cathode to the 2.7-megohm resistor going to the focus control, and the anode to the gray wire from the flyback; this went to pin 9 of the 1V2 tube. Mount this on a small terminal strip soldered to the frame of the flyback. You can put a dab of high-voltage putty around the terminal for luck, if you want to.

Chapter 4 CRT Troubles: Flyback, Focus, Voltages, Etc.

Aristotle was an ancient Greek philosopher—one of a group that left us some valuable ideas. Their personal habits would get them an X rating now, but their other ideas were pretty good. Aristotle gave us a two-valued system of logic. This hung on for quite a while. Everything was either black or white, A or B, on or off—nothing in between. Sometimes this is good, sometimes bad. At any rate, it makes a good illustration of methods for dealing with one problem that has become a lot more common lately.

In far too many cases, we have a tendency to assume that only one thing can cause a given trouble. We overlook possible alternative causes. This particular problem was very rare; however, it is showing up more and more these days. So I thought it would be a very good idea to take a look at other solutions.

BIAS PROBLEMS

Here's the problem: a complete loss of raster. High voltage may be completely missing, or very low. Yet, when we check out the high-voltage supply, or change everything in it, we *still* have the same problem—no raster! So, the thing to do now is revise our original (and quite logical) conclusion, that there was trouble in the high-voltage supply, and look for *another* cause.

The other possible cause is a defect in the *bias supply* of the picture tube! If the picture-tube bias isn't correct, the big bottle can draw so much *current* that it loads down the high voltage to the point where it disappears—a whopping 2 mA, or something in that area! That doesn't sound like a lot; however, *maximum* rated current for a standard 3-gun color picture tube is 1600 μA—1.6 mA! In terms of power, where we multiply voltage by current, using 25 kV for E, we have a very respectable amount of load indeed, at minute currents. So our high-voltage supply simply collapses under such a load, and the raster goes out.

There are several things that can cause this condition. Anything that upsets the dc voltages on the video amplifier will vary the picture-tube cathode voltages; upset the dc voltages on the three color-difference amplifiers, and you vary the voltage on the grids of the picture tube. This refers to the original circuit. The later RGB circuitry is subject to the same kind of troubles, though with different parts causing it.

In the original circuitry, we run the cathode at around +300V and grids at around +200V (ballpark figures, of course). It doesn't take much of a change in the right direction to cause trouble. If its grid is made more positive, it draws more plate current.

In this circuit, we have two possible variables; if we make the grids more positive, up goes the beam current. Also, if we make the cathodes *less* positive, we are actually making the grids more positive, since the grid voltage is referred to the cathode voltage! The actual bias is the *difference* between the grid and cathode voltages, regardless of what these voltages read to ground!

Let's assume that something upsets the bias on all three of the difference amplifiers at once. (In many circuits, the horizontal *blanker* can do this.) Suppose this causes their grids to go highly negative. Here, too, we can have a cathode problem; the blanking is fed to the common cathode; make the cathodes *more positive*, and you've made the grids more *negative*. So the difference-amplifier tubes are cut off; their plate current goes down and, naturally, their plate *voltage* "goes up" (more positive). So does the grid voltages of the picture tube, since these are directly coupled to do the difference-amplifier plates. Up goes the beam current of the picture tube and, in many cases, out goes the raster.

We can get the same thing in the video amplifier stage. Let's have a fault, such as a short in the video amplifier tube that makes it draw very high current. Its plate voltage goes down or more negative; so, the picture tube cathodes follow, leaving the picture-tube grids much more positive! Same result. An open in the "video peaking transformer" (used in a great many video output circuits) can even leave the video amplifier with zero plate voltage. Very bad!

In some cases, where the fault isn't quite as severe as these, you will get a smaller change in picture-tube bias. This will cause a "flare" symptom. The picture will get very bright, and the brightness control will have no effect on it. If the high-voltage supply happens to be in very good shape with a high reserve power rating, this can happen—in a few instances, with a small amount of raster pull-in on the sides. These seem to be contradictory symptoms, but looked at in this light, no.

The same thing can show up in the later model sets using RGB circuitry. In these, the grids are common, and generally tied together. Video and color signals all go to the three separate cathodes. But you can still have this type of trouble. It may be from a different part, but the same thing can happen.

This is the major problem that seems to be getting the boys—they don't remember this! As I said before, this kind of trouble used to be quite rare. I have run into at least a dozen on my own bench this year! (I don't explain 'em, I just tell 'em.)

There's a quick check for this, thank goodness. When you have made the initial tests and substitutions (the easy ones) with no results, just pull the base off the picture tube and recheck for high voltage. In sets with series heaters, you'll have to pull the high-voltage lead off the tube. If the high voltage jumps back up to normal, there you are. Go thou and check all of the bias voltages on the base of that picture tube!

The worst problem seems to be that we don't *suspect* this; or, we just plain don't remember it! There is always more than one possible cause for any kind of trouble. If you don't dig it out in the high-voltage supply itself, go chase some biases.

HEATER-TO-CATHODE SHORTS

Heater–cathode shorts in color picture tubes are the customer's pet catastrophe. With the price of pin tubes today, you can see why. The typical symptom of this disease is a raster obviously and uncontrollably all one color: red, green, or blue. The shorted gun has lost its bias and usually runs wide open, to full conduction.

However, there's an easy cure: same old principle we've used in black & white tubes for years. Just clear the short by isolating the heater winding from ground. We used to do this with "brighteners" in black & white sets; however, heavy-duty types to handle the higher heater current of color tubes weren't available until lately. Now they are, from Perma-Power and others.

Figure 4-1 shows what happens. The heater of the color crt is "dc-biased" to reduce the voltage stress between cathode and heater. Color-tube cathodes run at about 150–200V above ground. So set makers connect a voltage divider between B+ and ground, and apply about 150V dc to the heater. This winding is not grounded like the others, of course. (This same

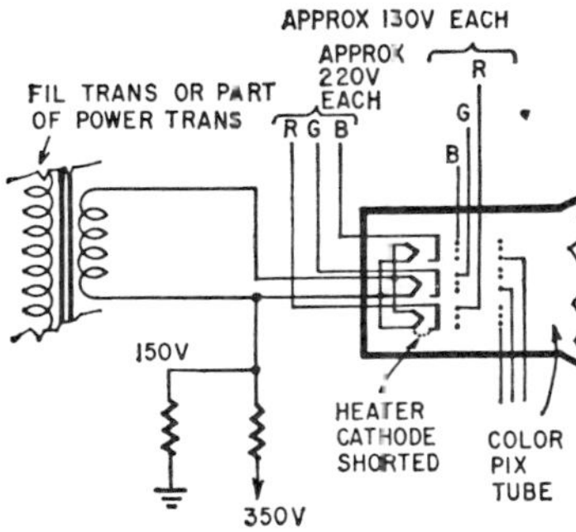

Fig. 4-1. Even thought the pix tube cathodes operate with low vootage, they're supplied their driving voltage well above ground. This provides "stress relief" by lowering the potential differences inside the tube.

biased winding normally supplies the damper or high-voltage regulator tubes, for the same reason—and, in one well known chassis, the second i-f tube!)

A heater–cathode short in the crt usually kills the dc bias, letting that gun go wide open, or kills the video (Y) signal, or both. *Don't* try to blow this type of short out with your crt tester rejuvenator; you'll usually blow the heater itself. The easiest way to get rid of it is to "lift" the heater from ground—isolate it.

The equipment needed for this is extremely complicated (Fig. 4-2)! It's a plain 1:1 transformer: 6V in, 6V out. This is not a brightener, though it looks like one on the outside. There is no increase in the heater voltage. All the transformer does is isolate ground from heater supply.

Even though this "ground" may be through the B+ supply lines, it will still be at a low impedance for the high-frequency

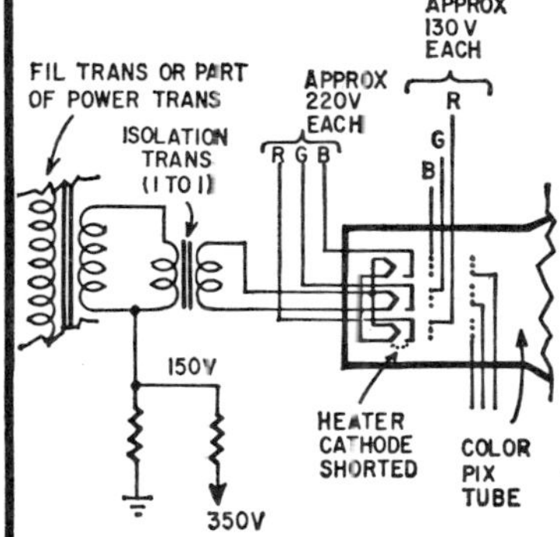

Fig. 4-2. Using the 1:1 transformer to isolate crt heater from ground.

video signal (through filter capacitors, etc.). It will also have the inevitable effect on the dc bias of the crt, also a problem.

Like the brighteners, the transformers are plug-in. Though the transformers are the same, you'll need two types: one for the older round-tube bases, and one for the rectangular types with the small base.

All you need on this job are a good crt tester to confirm your diagnosis of a shorted crt and a pair of these isolation transformers.

We've seen tubes equipped with similar devices last for several years. Before the introduction of these, we had to use separate filament transformers mounted in the cabinet. The modern method is a lot simpler and easier.

FOCUSING

There are a few things which ought to be cleared up on the general subject of focusing of a color picture tube. These include what it is, what it does, and what voltages you should expect to read. There are also a few common beliefs that aren't true: we'll see about those too while we're here.

Focus Voltage

The focus voltage in a standard tri-gun picture tube is not necessarily a *fixed* value. That is, you can't say "I'll set the focus voltage at exactly 4628 volts and that's it!" The focus voltage must be a *percentage* of the high voltage on the ultor at any given instant. This will be roughly 17–20% of the high voltage. For 25 kV, focus will be very close to 5000 volts. If the high voltage drops, the raster will defocus if the focus does not drop with it!

Fortunately for us, the design of the high-voltage supply, in almost all color sets, is such that the focus voltage and the high voltage *will* rise and fall together. They both come from the same source. The high voltage is developed by the high-amplitude flyback pulse coming back from the yoke during the horizontal retrace interval, or flyback time. This is stepped up because the high-voltage winding is coupled to the primary of the flyback. Peak voltage will be about 30 kV. This is rectified and fed to the ultor of the picture tube.

If you check the schematic of a flyback, you'll see that the focus rectifier is fed directly from this pulse. It is tied to the same connection as the plate of the horizontal output tube, on the primary of the flyback. Typical pulse amplitude at this point is about 5000V peak. This is connected to a small high-voltage rectifier, filtered just a little, then applied to the G3 or focus grid of the picture tube. This isn't actually a grid.

There are three little cylinders, one in each gun, all connected together.

Like a grid, though, this is a single-ended circuit. It goes nowhere; just sticks out inside the tube and stops. Now; here's a valuable fact which can be used in analyzing many focus problems. This is absolutely a dry circuit; no current flows in it at all. This is why you find very high-value resistors in series with the focus supply. Since there's no current flow, there's no voltage drop across them.

If you do see a voltage drop, the chances are that this is a meter drop due to the loading effect of your voltmeter. High-impedance voltmeters, with high-voltage probes, should always be used for any readings in these circuits.

Focus Current

Under some fault conditions, you will find current being drawn by the focus circuit. This causes a large voltage drop across the big series resistors, a drop in focus voltage at the socket, and a severe defocusing of the raster. The main cause is gas in the picture tube. Too many of us call this grid current, but it's really gas current. In any case, there's a quick check.

Pull the picture tube socket off the tube. Now, recheck the focus voltage at the socket pin. If it has now come back up to normal, and the focus control will cause a normal variation (about 2000V), the focus circuit is okay but you need a new picture tube! If the focus voltage does not come back up with the picture tube disconnected, *then* you check the focus supply circuit.

WHAT IS FOCUS?

The focus voltage does exactly one thing: it focuses the *horizontal scanning lines* in the raster! Period. It has absoblastedlutely *nothing* to do with the horizontal resolution of the picture, and darn little with the vertical resolution.

Defocusing will cause a slight smear of *all* objects in the picture, if it's minor. If it's bad enough (1000V low, say) all objects will be badly smeared. A complete loss of the focus voltage will usually put the raster completely out. If you find a dark screen, but the high voltage is up to normal, read the focus voltage just for luck. If you can adjust the focus control, and get good sharp scanning lines on a blank raster, the focus voltage is normal!

One common complaint is "this set has focus problems! The closeups are sharp, but things in the background are always *fuzzy*!" Check the scanning lines. If these are sharp, your focus is good—most especially, if *closeups* are sharp.

This phenomenon is perfectly *normal.* It's due to the aperture being used on the lens of the *TV camera.*

Check any photograph with a clear sharp image of something in the center. Now check the background; it will be fuzzy and "out of focus." A lens will not focus over an *infinite* distance. When you "stop down" or make the aperture smaller, the depth of focus (depth of field) will be longer, but you get much less light. Beside this, they generally don't want the background anyhow. What they're interested in is the person or thing in the closeup.

Focus Rectifier Problems

The original color sets used small tubes as focus rectifiers, such as the 1V2 and others. The new sets use solid-state focus rectifiers.

Tube rectifiers can get weak from old age, like any other tube. This is easy to check; just pop in a new tube. If the focus voltage comes back up, this was it. The solid-state focus rectifiers can also "grow weak." If enough of the little pills short or leak, the output voltage will drop. If they arc over internally, they'll sometimes make the case bulge or blow open in the middle.

In general, a solid-state focus rectifier can be substituted directly for any of the tube types. This is a very handy thing, especially in those irritating cases where the socket of the focus rectifier has arced over and burned up. Instead of replacing the socket, which is usually a rough job, take the plate lead loose. Mount a solid-state focus rectifier on terminal strips to keep it from arcing to other things, and tie the plate lead to the anode end. Tie the lead from the focus control to the cathode end. Be sure to clean up and tie the heater leads of the original tube, so they can't get into any trouble.

There are lots of readily available solid-state replacements; RCA SK-3066, and others. Watch the voltage, though, there's less drop in a solid-state rectifier, and you'll wind up with higher output potential.

Final hint: in a dry circuit, you would think that any series resistance would have little effect. Yet, if the focus pin or its socket contact on the picture tube "corrodes," this can cause a very tricky *intermittent* loss of focus! Clue: focus voltage, read at the focus control, source, etc. does *not* change when the raster defocuses.

To check for this, pull the picture-tube socket off and examine the focus pin and the socket contact. This will be the one with a pin missing on each side of it. If you can see distinct traces of a light green powdery substance, look out! You may find that the socket contact has been almost eaten away. This is apparently the same as the green "corrosion" we used to see in audio transformers. etc. in old radios. If it's bad enough, the socket must be replaced.

FLYBACK TRANSFORMERS

There is nothing more embarrassing than replacing a flyback, turning the set on, and finding exactly the same old symptoms you had before. Besides being expensive, this kind of thing has a tendency to make you doubt your ability to make a living at your chosen profession. So we need to be *sure* before we take that last step.

There are several tests which will give a good indication of the condition of a flyback. Notice that I said *good*—not *definite*. There is only one absolutely definite test for a shorted flyback—this is done with the *well calibrated eyeball*, and consists of observing a thick layer of soot on the inside of the cage, and the flyback itself lying in a charred lump in the bottom.

Load Loops

The worst faults are the ones which indicate that there is a short in a flyback, when the actual short is *external*. There are a few of these in a black & white TV set—in color sets, quite a few more. A flyback/yoke circuit is actually a *tuned circuit*. Like a radio transmitter, anything which draws energy from this circuit is a *load*. If one of these normal loads is shorted, it will draw excessive energy from the flyback, cause a very high load current to flow and, in time, overheat the transformer and make it burn up.

The quickest single test to show up an overload is the cathode current of the horizontal output tube or transistor. This stage *controls* all of the *power* used in this circuit (doesn't *supply* it; it comes from the low-voltage or B+ supply, but it does control it). If there is even a small overload in *any* of the load circuits, it shows up instantly as an increase in the cathode current of the output stage.

There is your key symptom. From this point on, your job is to find out what is causing this excess current to be drawn. If the flyback does have shorted turns, you'll get this kind of symptom. However, there are quite a few things which can "fake you out" on this, so watch out for them. We'll assume that the B+ supply is normal, the tube or transistor is good, and that the input drive signal is normal. These are easy to check. So are things like a shorted boost capacitor, and so on. The object of these tests, from here on, is to *make sure* that the flyback actually is internally shorted, or isn't.

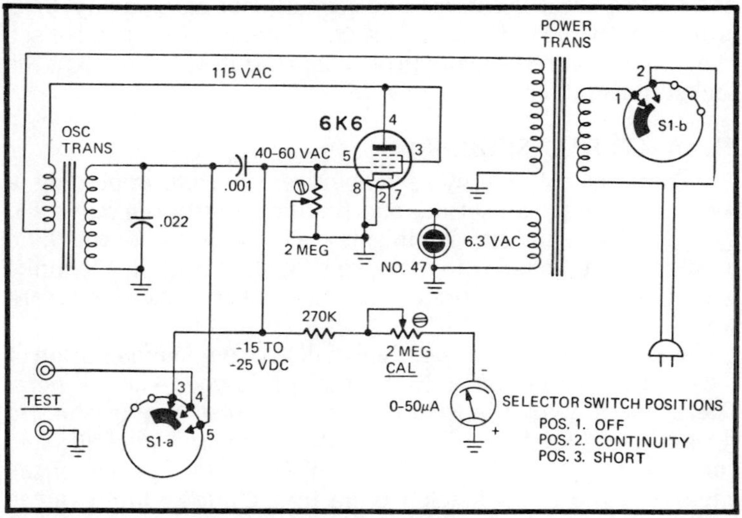

Fig. 4-3. This circuit is the time-honored flybacker—a 16 kHz oscillator that lets you use a suspect flyback transformer as an inductance in the circuit. Shorted flyback-transformer turns lower the circuit Q and change the meter indication. (Meter monitors grid current.)

Test Instruments

One essential tool for this is a cathode-breaker adapter, to let you read the cathode current of the output tube during operating tests (Pomona 2599-A for 6JE6's and any Novar tube with the cathode on pin 3, etc.). For actually testing the flyback for shorted turns, my pet test instrument is a very old design, the "flybacker." This was made by Radio City Products Co. a long time ago, and is still being built by several companies. Figure 4-3 is the schematic of a typical unit.

This is nothing but a 16 kHz oscillator. The meter reads the grid current. If a good inductance is connected across the test leads, it will not damp the oscillations, and the meter stays in the GOOD sector. If the inductance has only *one* shorted turn, it will reduce the Q of the whole circuit so badly that the oscillation will be reduced, and the meter will fall and read in the REPLACE sector. This simple unit is amazingly accurate. You can verify this: connect a good flyback to it, note the reading, then wrap one turn of wire or solder around the core. Short the ends of this, and watch the meter fall into REPLACE. This makes a "one shorted turn" loop coupled to the inductance, and pulls the Q down.

However, you *and* the flybacker can be fooled if there is too much external loading on the transformer—especially in color sets where many circuits are connected to the flyback: pulse circuits, the high-voltage rectifier, focus rectifier, and so on. The filament circuits of the rectifier tubes make up a closed loop, and can show fake loading.

Now let's run through the whole procedure to see how much of it we need to use in any given case. The dc voltage supply has been checked, drive checked, tubes checked. There is no normal output (sufficient high voltage, sweep, and so on) and the cathode current meter shows a reading far above normal. Ohmmeter tests around the boost circuit, etc. show that there are no nice easy shorts to ground. So, at this point that there *is* a distinct possibility that the flyback is shorted.

We pull the high-voltage rectifier tubes (to open those closed loops), and take the cap off the horizontal output tube. Connect the *flybacker* between the plate caps of the output tube and high-voltage rectifier. It says SHORT. There you are. However, *if* you stop testing at this point and order a new flyback, you may be in deep trouble. All of the returns aren't in yet.

There is one thing that we *must* do, before making a definite decision that a flyback is shorted. Make sure that *all* external loads have been taken off. This is not too hard to do, and it can save you a lot of time. Check the schematic to see how many normal loads are used. There won't be too many.

First, disconnect the yoke. Next, take off anything like a width coil, which is connected in shunt across a part of the flyback windings. In many color sets, you'll find one of the focus transformer windings connected directly across the bottom of the flyback. While doing this, leave the *flyback* hooked up to the plate caps of the output tube and high-voltage rectifier.

Each time you unhook a wire, watch the meter. If there's no change, keep on going. If you get *all* the wires off the flyback, and it is still in the SHORT sector, go ahead and order a new flyback, However, if your short is in one of the load loops, you'll see the meter jump happily back up into the GOOD sector. Figure 4-4 shows a partial schematic of a typical flyback, with the different load loops indicated.

Case Histories

Let's look at a few actual "fake short" cases. One of them was a Magnavox 920 color chassis. The 6JE6 was glowing red-hot, and there was no raster or high voltage. New tube didn't help; cathode current was 400 mA plus (that's as far as my meter went). The *flybacker* showed a short, when connected to the 6JE6 and 3A3 plate caps. Disconnecting the

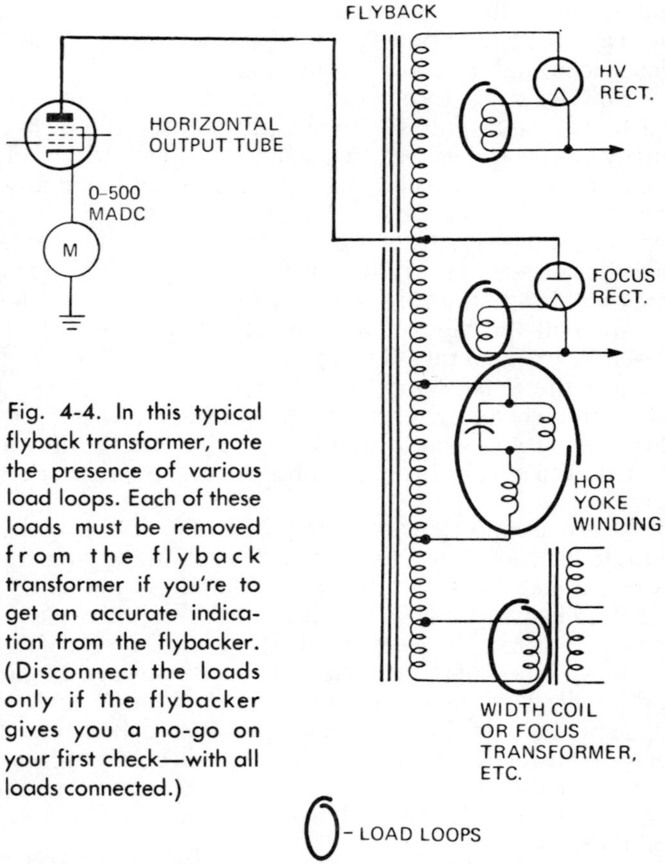

Fig. 4-4. In this typical flyback transformer, note the presence of various load loops. Each of these loads must be removed from the flyback transformer if you're to get an accurate indication from the flybacker. (Disconnect the loads only if the flybacker gives you a no-go on your first check—with all loads connected.)

yoke didn't help; it stayed there until I took off the leads going to the focus transformer winding. This eliminated the short; the meter went upscale. It will do this when connected to a good high-Q flyback, which most color sets use. That was that; the focus transformer was shorted and the flyback was saved.

In an RCA CTC 22 portable color set, same thing. Flyback showed short, and all of the typical symptoms were present; high cathode current, no high voltage, etc. This one was caused by a shorted damper diode, which in this case was a solid-state type under the chassis. In similar circuits, a shorted boost capacitor, or the capacitor connected across the solid-state damper will show the same thing.

In a very small GE hybrid black & white TV chassis, there was no high voltage. In-circuit tests showed the flyback shorted. This one was mounted directly to a PC board. After taking the flyback out with a desoldering iron, it checked perfectly good. Here the short was in the horizontal yoke winding, another dandy load loop.

Summary

Once the high cathode current of the horizontal output tube, together with the lack of "normal output" (no high voltage, no horizontal sweep, etc.) , has shown you that there is a possibility that the flyback is shorted, make the necessary tests. **But don't stop short of making all of them!** Take off every external load and repeat the test for shorts. Even in a color set, this won't mean disconnecting more than about seven or eight wires. Leaving the *flyback* hooked up and checking its reading each time you take off a wire will give you an "instant clearance" on the flyback, or a defininte indication that it is bad.

The only place you can get into trouble with this type of instrument is with very low-inductance *yokes*. These will normally read very low—in some cases into the SHORT sector. If there is a doubt, take the cap off the back, and check each half of the choke separately. If these readings are the same, the chances are that the yoke is good. Very few yokes show identical shorts in *both* halves.

READER QUESTIONS

The questions presented in this subsection generally relate to problems associated with crt circuitry, flyback transformers and yokes, focus problems that might be attributable to defects in the crt circuitry, and the like. Also included are many questions pertaining to areas not in the above-mentioned areas but which might have the appearance of being traced to these functional elements.

If you have a specific problem that you feel might have been asked in the Service Clinic, but you don't find it in this subsection, also check the questions presented in other chapters.

Focus Trouble

This Philco 18QT85 has a focus problem. The focus coil doesn't seem to react at all. Focus is poor, although everything else seems to be okay. Can you tell what's causing this?—V. T., San Leandro, CA.

We'll get close. If your boost voltage and width are normal, and high voltage is close to normal, then the focus circuit has the normal supply. This is the 5 kV pulse developed

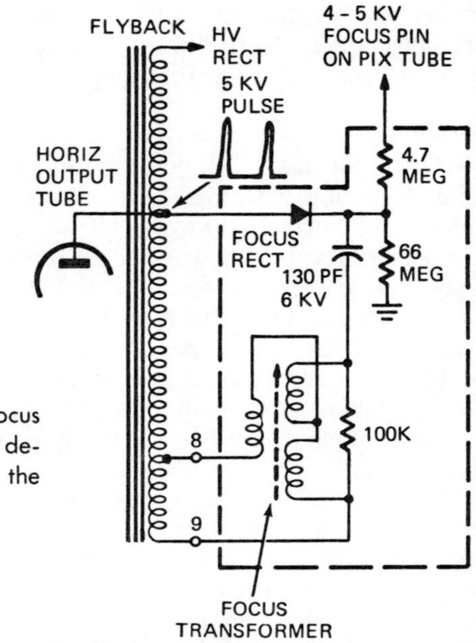

Fig. 4-5. Trouble in focus circuit may be due to defect in a part within the dashed lines.

at the plate of the horizontal output tube. So your trouble will be in one of the 5 parts inside the dashed line in the sketch (Fig. 4-5).

If your focus voltage is low, try a new focus rectifier. If the focus voltage is high and the focus adjustment has no effect at all, the focus transformer may be open. (If it is shorted, it will load the flyback very heavily.) A focus voltage that is too *high* or too *low* will cause defocusing.

Weak Video

With a new picture tube in this Silvertone 8175, convergence, color, and so on are good, but the video is very weak, apparently. The picture is clear, but pale and washed-out looking. Agc works, but won't clear up the problem.—K K., Skokie, IL.

This has a very familiar ring to it. Check the 60 μF electrolytic capacitor on the cathode of the 12BY7 video output tube. Note that the contrast control does *not* vary the cathode resistance, but only the position of this capacitor on that resistance. (See Fig. 4-6.) So it varies the amount of degenerative feedback from the cathode circuit. If the capacitor is open, you'll have maximum *feedback* at all times and symptoms exactly like those you describe.

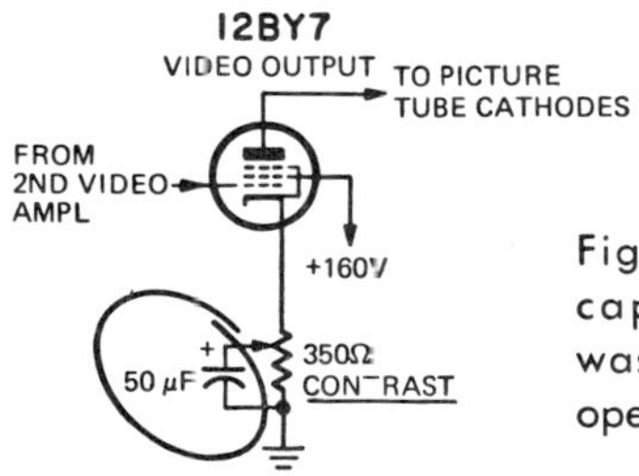

Fig. 4-6. Electrolytic bypass capacitor may be cause of washed-out pix. It's probably open.

Picture Disappears

The picture will disappear on this RCA CTC 17X. It goes so quickly that I can't tell whether it's blooming or just fading out. After the raster goes dark, I can see a couple of faint lines at the left side. These weave and bend. The high voltage stays up pretty well.—W.R., Spokane, WA.

If you're losing the raster, but the high voltage stays up to, say, not less than 20 kV, then the root cause of this would be something in the picture tube *biases*. In other words, the tube is simply being cut off, either by too much positive voltage on the cathode or too much negative voltage on the grids. Monitor both of these, and see which one changes when the picture goes out. You'll probably see a change of about 50V, to cut the tube off completely. Check the blanking circuitry, too, and the kine-bias control.

Picture Tube Heater Dead

This Emerson 129021 "Montclair" has a series heater string. All other tubes are lit, but the picture tube heater is dead. The tube tests okay.—R. R., De Queen, AR.

The picture-tube heater in this set is fed from a small transformer; this is for the quick-on circuit. The primary of this transformer (Fig. 4-7) has a diode and a nonpolarized electrolytic capacitor in series as shown in the diagram. Check for ac voltage across the transformer primary. If you don't get the normal 113V ac, either the diode or the capacitor is open.

Four Black Bars

The screen of this Zenith 20Y1C48 shows four black bars, covering the right ²/₃ of the screen. about 3 inches apart—D. W., Rochester, NY.

The most common cause of this kind of trouble in these sets is the horizontal blanking diode. If it gets slightly leaky or shorts, it will let the blanking pulse and the ringing along the

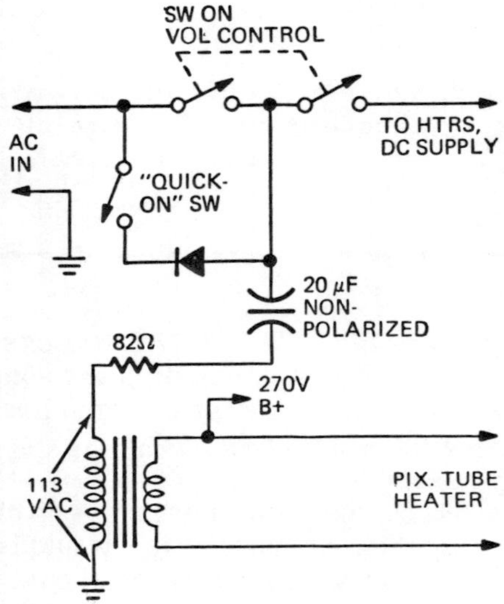

Fig. 4-7. The diode or capacitor in this set will interrupt crt heater voltage, if open. Most likely culprit is the back-to-back electrolytics (they make a nonpolar capacitor). If the capacitors aren't identical, one will display its polar characteristics—which results in punchthrough when ac is involved.

baseline get through. It's not supposed to let that baseline get by.

This is a tiny diode, located under the chassis, in the center of a triangle formed by the 6EJ7 burst amplifier, 6JU8, and 6KT8 second bandpass amplifier tubes. Take it loose and check it; better still, try a new one.

There's one other possible cause for this, but let's hope it's just the diode; the other one requires a new flyback. (Usually makes *whites* lines, if this is it.)

No Boosted Boost

The boosted boost is very low on this Heath GR-295 color TV. I'm only reading about +700V instead of +1000V. I've tried new rectifiers and capacitors. R718 (Fig. 4-8) burned out; I replaced it and the new one gets hot and drops in value. Think this might be the flyback?—F.B., Lemoore, CA.

Nope. Look at your major symptoms. You do have *normal input* to this circuit: the boost (with pulse). If you did not, you wouldn't have enough power to burn up that resistor. So, from the symptoms, and the fact that the rectifiers and capacitors have been (rightly) replaced (see diagram), you've got an *overload* on the 1000V line. *This* is burning up the 470-ohm

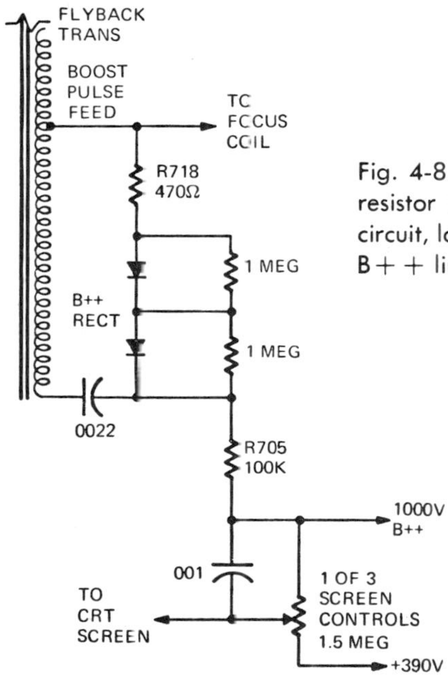

Fig. 4-8. When low-value flyback resistor burns up in this type of circuit, look for an overload on the B++ line.

resistor! Normally, there is very little current flow in this circuit.

Check the bypass capacitors on the screen controls, and all wiring on the 1000V line. These are the only loads on it.

No Raster

After I fixed some other things, I had this GE CB chassis working very well. Put it back in the cabinet, and it won't light up! Plenty of HV. Voltage on 12HG7 video output plate is about 50V high. Won't change as the brightness control is turned. Where to look?—H. S., Vincennes, IN.

Right where you are now! That excessive 50V on the 12HG7 plate is normally enough to bias off the picture tube cathode and kill the raster! Trace the circuits of this tube; check to see if the cathode circuit has been accidentally opened as you were replacing it in the cabinet.

Crawling Cathode Current

This Sears 528.81201 has had me in a spin for days. The 40KD6 cathode current slowly rises from 210 mA to over 400 mA. High voltage rises to 30 kV, regulation won't work, and finally the breaker trips. Grid drive remains normal at 200V p-p, but the grid voltage goes from −62V to −42V. All tubes, flyback, etc. have been changed. No help. Does this with picture tube second anode unhooked, too. Any ideas?—D.B., Bremerton, WA.

The crystall ball tells me that this must be something related to that change in grid bias on the 40KD6. This drop in bias would let the tube "stay on" too long, thus making the output *rise*, and the current rise with it.

I'd suspect the high-voltage adjust circuitry, which in this set is a grid bias control, with a vdr, fed by a pulse from the flyback. It *should* hold the output and current *down*, but it seems to be working backward. It is supposed to develop a higher *negative* voltage if the output tries to go above normal. However, you're getting a *less negative* voltage on the 40KD6 grid. Check the bias across that vdr, also the pulse height. I don't have the exact figure, but this should develop about −100V at this point.

Plenty Voltage, No Focus

I have sat here and glared at this RCA CTC 28 chassis until my eyes hurt! I can't get anything that looks like a focused raster! Yet, I've got 25 kV of HV, and the focus voltage will vary up and down from about 5200V to 5400V. Every part good. Tried a new 4.7-megohm resistor just for luck, although I was reading the focus voltage at the load end of the thing.—T. T., Ink, AR.

Cheer up. I've got eyestrain and brain strain too from the same thing. Pull the picture tube socket off and look closely into the hole for pin 9. If you see a whitish or pale-greenish dusty looking deposit there, that's it. The 9-pin of the tube will probably show the same kind of thing. Also, if you pull gently on the black wire, it will probably come out of the socket! The focus pin on the crt socket *corrodes.*

You wind up with a very bad contact, and in some cases, the focus pin of the socket is eaten entirely away. If you have a taste for knick-knacks, you can cut up a standard 7-pin wafer socket out of an old radio or something, take one of the socket contacts, and make a new one. You'll have to drill out the two rivets that hold the socket body together and replace them with No. 2 screws when you put it back together. (It's a lot easier to replace the crt socket; I've tried both ways.)

Checklists For Dim Raster

The raster on this Magnavox 904 is very dim. What should I check to find out what's causing it?—W. B., Philadelphia, PA.

1. High voltage—23–25 kV minimum.
2. Focus voltage—20% of high voltage.
3. Pix tube cathode voltage.
4. Pix tube grid voltages. (The difference between these is the actual bias on the pix tube.)
5. The pix tube itself, for emission. Must have a minimum emission of 500 μA per gun for good brightness.

If the video output tube's plate voltage is low, or the color-difference amplifier tube's plate voltage is high, this will change the crt bias to the point where the picture tube cannot conduct enough beam current. Faults in these little tubes can cut off the crt tube.

Red Bloom; No Focus

I can't get good focus in a Sylvania DO-9 chassis, even with the core of the focus coil all the way out. It seems that the blue and green focus pretty well, but the red doesn't. Checking with magnifying glass shows that the red dots are much bigger than the others.

The focus voltage on pin 9 varies from 4000 to 5000 volts, with high voltage of 25 kV. Grid and cathode voltages on the pix tube look okay: About equal on all guns. Does this look like a bad picture tube?—J.A., Dallas, TX.

Not necessarily. There are a couple of things which can cause this symptom. Doublecheck: Turn off the red screen and see if there are clear scanning lines on the blue-green raster, or either one by itself. Then recheck the red raster all by itself. If you can't see the scanning lines now, than you *do* have a red bloom problem in this color set.

Sylvania's "troublehsooting schematic," a part of which is shown in Fig. 4-9, shows one possible cause for a single-color bloom: A shorted bypass capacitor across its screen control. If the capacitor were only leaky, it might account for the slight bloom you have on the red.

I *think* your focus voltage variation, with the HV you have is normal. Focus voltage should be about 20% of the existing HV. The 5000V maximum you can reach should be enough to get good focus with 25 kV of HV. If you can focus *any one* raster, of course, then the focus voltage itself is okay.

Drive Lines

I have a Magnavox solid-state TV chassis with a very peculiar set of what looks like drive lines on the screen. One is where you'd expect it, about the center, but the other is way

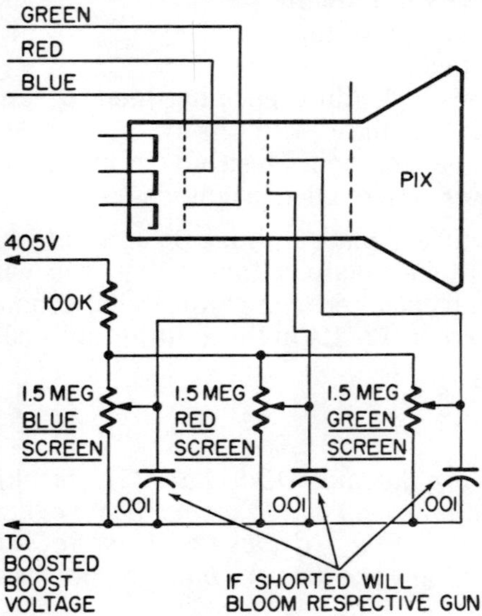

Fig. 4-9. When only one color blooms (Sylvania DO-9), look for a bad bypass capacitor at the respective screen control.

over to the left. There is some small ringing evident in the raster at that side.

Drive waveforms all look fairly close to those shown on the schematic. HV up to normal. What is causing this?—E. K., Media, PA.

This is yoke ringing, really, despite the fact that the symptoms do resemble "drive" lines like those we found in tube TV circuits when we had trouble in the horizontal output tube grid circuitry! In this transistorized horizontal output circuit, the yoke is driven by a sharp square-topped pulse.

To prevent ringing, there are two small capacitors shunted directly across the horizontal yoke windings, which are connected in parallel. One is a feedthrough 0.001 μF, near the horizontal output transistor on the chassis, and the other a 0.0082 μF connected to pin 7 of the yoke socket. If the 0.0082 μF capacitor is open, you'll get exactly this kind of symptom. Check it by tacking another in place of the original (with the power off; don't bridge capacitors in transistor circuits with power on!).

Picture Tube Heater Won't Light

The heaters of the picture tube won't light in this RCA CTC 38. The picture tube is good on a tester. The power transformer gets pretty hot. Do you think the transformer is shorted?—J. G., Mena, AR.

Not yet. Take the 6.3V heater leads from the power transformer loose; the picture tube heaters are fed from a "dc biased" winding on the power transformer. A bias of +195V is connected to this heater (Fig. 4-10) to reduce the chance of heater-to-cathode shorts. These are the green and green-yellow wires. Now turn it on and read the ac voltage across this winding.

If it is okay, check the 6BK4. Its heater is also fed from the same winding. In a few cases, this tube has shorted and overloaded the picture-tube heater winding. This will also make the transformer run very hot. While you're at it, look out for accidental wiring shorts across this winding. In a few models, the second i-f tube heaters are also connected here.

Burned Out Flyback

The flyback has burnt out in this Admiral 4G13 color chassis. There are two horizontal output tubes in parallel. I'd like to protect the new flyback against tube shorts. No fuses used that I can see. How would I do this!—C.G., Mountain Top, PA.

Open the cathodes, tie them together, and add about a ⅜A fast-blow fuse to ground. Don't use slow-blow fuses; they have enough resistance to cause degeneration. Beside that, you want that circuit to open as quickly as possible if anything happens!

Trapezoid Raster

I've never seen a case like this! The raster on this Olympic CT-910 is distorted, as in the drawing (Fig. 4-11). The scan lines pull off and up to the left in the top left corner of the screen. If I adjust controls to make the raster smaller, it shows

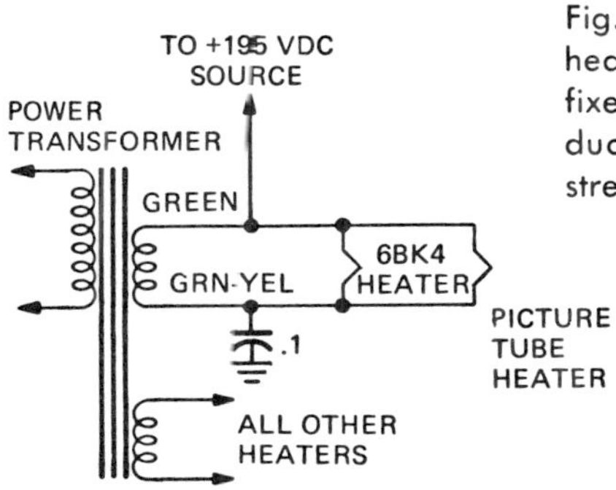

Fig. 4-10. Picture tube heater is biased by a fixed dc voltage to reduce heater—cathode stress.

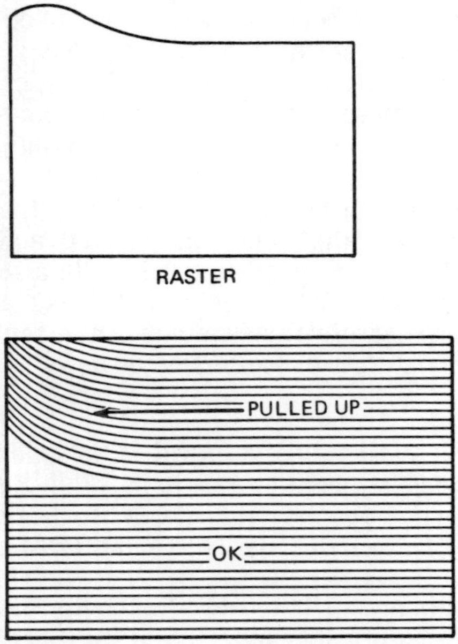

Fig. 4-11. Raster pulling like this is rare; but when it happens, the only cure is a new pix tube. The trouble: an open at the ultor contact. Wipe the ultor area clean and examine the connection with room lights off; arcing will show.

the same shape. Varies with brightness.—J. H., Pittsburgh, PA.

This is a rare one. I've seen it happen only twice (once on *my own* set!). This problem is almost certainly due to an *open connection* inside the picture tube—at the ultor button. The inside coating of Aquadag has broken loose. The high voltage arcs across the gap and apparently creates a magnetic field which causes the pulling up of the raster lines.

Wipe the glass clean around the ultor connection. Now darken the room, and turn the set on. If this is the trouble, you'll see tiny arcs, very obviously *inside* the glass, usually all the way around the ultor button. A new tube is the only cure.

Blackout

Here's one for you. A GE M587 had no raster. Pulling the audio tube restored it to normal. A check showed 120V on the picture-tube cathode, but only 15V on the control grid. So the tube was cut off. Capacitor C201 (Fig. 4-12) was leaking, and the result was the drop in grid voltage which cut off the tube. —James Franks, Verdun, Quebec.

Jim Franks told us about a GE that did not display a raster until the audio output tube was pulled. The basic trouble was due to a shorted 0.01 μF capacitor on the blanking network.

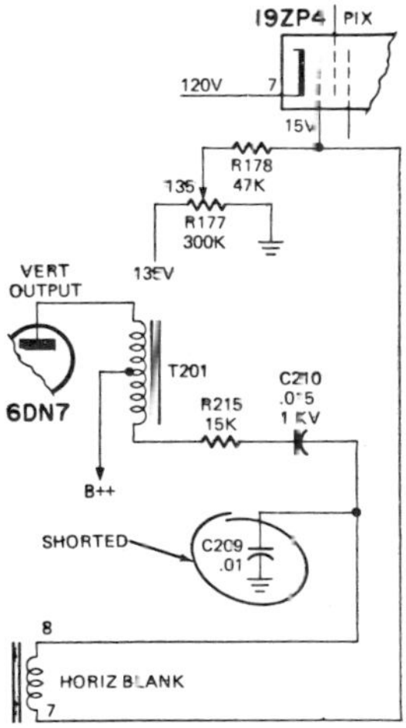

Fig. 4-12. Reader submitted this drawing with question. See response.

The accompanying diagram (Fig. 4-12) did not include the audio output circuit, so the connection between the blackout and the audio circuit was not immediately obvious.

This TV receiver uses a stacked B+ supply with the cathode of the audio output serving as the 135V B+ source—supplying, among other circuits, the cathode and grid of the picture tube (see Fig. 4-13).

Normally, the cathode is biased at a fixed value of 120 to 130V and the control grid bias varies from 0 to 135V with the setting of the brightness control. When C209 shorted, it pulled the control-grid bias way down so the picture tube was cut off Pulling the audio output tube removed +135V as a source of bias for the picture tube. The bias voltages—now available from the 268V line through interconnecting resistor networks—are such that the picture tube is biased on, so the raster returns.

"Pulsing" Focus

I have a very peculiar problem with a Zenith 25MC33 color chassis. The thing "pulses" in and out of focus at a rate of about 2 or 3 seconds. The funny thing is that the focus voltage, read at the focus rectifier, does not go up and down as you'd expect it to!

I wiggled the damper tube and it cleared up momentarily; so I cleaned the damper socket contacts. Now it's back again!—C. H., Port Angeles, CA.

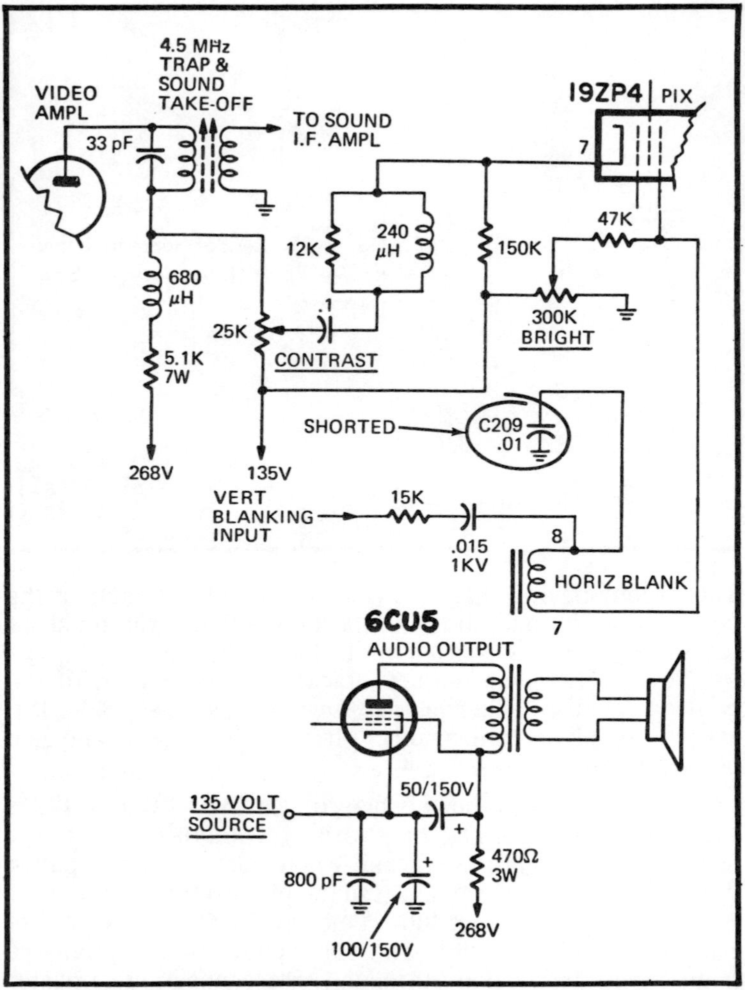

Fig. 4-13. In this circuit, the cathode of the audio output tube serves as a B+ source! A short in C209 pulled the bias down, thus cutting off the pix tube.

From your description of the symptoms and from some similar experiences on other makes, I have a suspicion. Take the crt socket off the tube and look at focus pin 9. Then check the No. 9 hole in the socket. If it is filled with a greenish powder or there are traces of it on the focus pin, look out!

Bare metal, charged with a very high voltage, and *exposed to the air*, will often "corrode." It is some kind of oxidation, and I don't understand the exact chemical reaction, but I'll guarantee that it *will* happen! (Used to eat wires in two in old radio output transformers, i-f's, etc.)

So take the socket apart. You are likely to find the focus pin contact eaten away or broken. The high voltage will "jump" the gap through the oxide, developing a varying resistance, making the focus change. Best cure is to replace the socket, although you can often take a contact from a small wafer-type 7- or 9-pin tube socket and make one.

Replacement Flyback For Old Stromberg

I need a flyback transformer for a Stromberg-Carlson TC-19 TV set. Don't seem to be able to find a replacement in any of my lists. Can you help?—J. H., Charleston, WV.

Not a heck of a lot, I'm sorry to say! I went through my extensive lists and came up empty too. However, I finally found a listing on this chassis, in a very old Thordarson-Meissner catalog! The replacement given for the S/C 16124 flyback was FLY-1.

By some extensive cross-referencing, I also found that a Triad D-11 will work. The original was referenced to the very popular (in them days!) RCA 211T1/211T3 flyback. Many, many sets used it, and there should be one floating around somewhere.

Odd Raster

After I changed the picture tube in this Truetone EIS2885A-86, I got sound but no picture. There was just a faint glow along the bottom of the screen. When I turned up the brightness control, the screen lit up from the bottom up, like a window-shade being raised. When the brightness control got all the way up, the screen flashed and went out. This repeats if I turn the brightness control all the way off and bring it up slowly.

Some voltages are weird; I read 4000V on the video output plate (Fig. 4-14), which should be +125V. With the picture tube socket off, I read −50V on 2, 6, and 11. High voltage reads 25 kV with the screen dark and only about 3 kV with brightness up. Need ideas!—M. F., Wichita, KS.

Here are a few: Check the continuity in the connections to the picture tube *cathodes*. That negative voltage you got with the base off is badly wrong. This should have been +150V or more. The dc voltage is fed to these cathodes and the video amplifier plate through a network of resistors and peaking coils. Frankly, it sounds very much as if the cathode disconnects have been fouled up or actually pulled loose. This would take the load off those resistors, and let the video

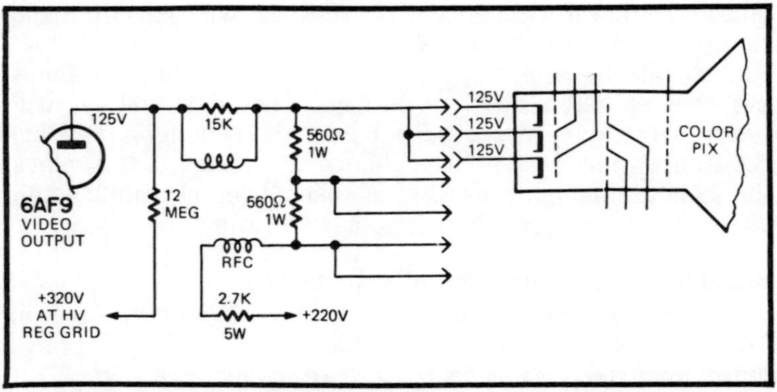

Fig. 4-14. Brightness control causes window-shade lighting effect on picture tube.

amplifier plate voltage go way up. Since it's fed from +220V, I don't know yet where your +400V comes from. However, if your cathodes are all at zero voltage, this *will* definitely cause the picture tube to draw a very heavy beam current and kill the raster by overloading the high-voltage supply.

No B+ Voltage On Damper Plate

This Admiral portable, a G4 chassis, has no raster. Sound is very low, but clear. I get no boost voltage at all; in fact, there's no voltage, even on the plate of the horizontal output tube! Also, no B+ on the plate of the damper tube! Help!—G. K., UT.

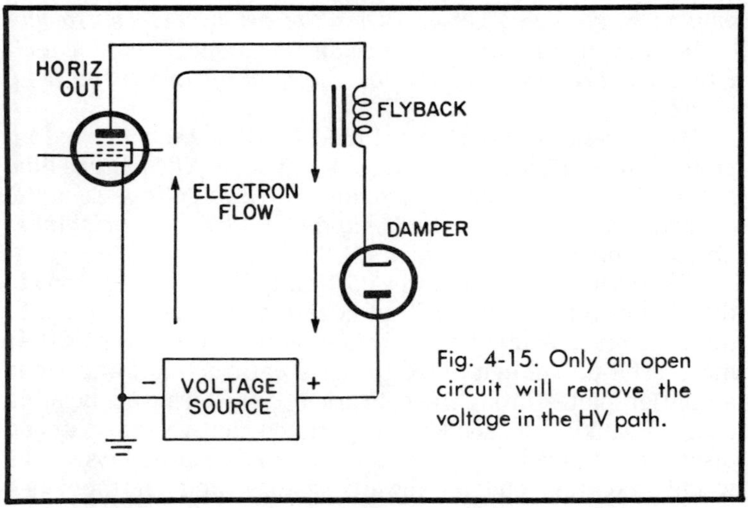

Fig. 4-15. Only an open circuit will remove the voltage in the HV path.

Ah yes! I've seen this one before. Start at the damper plate, pin 2 of the 38HE7 combination horizontal output/damper tube, and check continuity back toward the B+ 135V source (see Fig. 4-15). You'll find an open circuit somewhere!

The damper tube must have B+ before any circuit like this can work. The voltage supply for the horizontal output tube plate comes through the damper tube from B+. So get this first and the rest will probably work. Check the PC board under the 38HE7 socket. If C432, across the damper tub, has been bent over, it can arc to the PC board conductor, which connects pin 2 of the 38HE7 to the load end of the rf choke, L403 in Fig. 4-16. If this has happened, it'll be easy to see. Bend the capacitor up and away from the board and run a bead of solder along the conductor. Job over.

Streaks in Raster

I've a weird symptom on a Magnavox 919. There are tiny horizontal streaks and flashes in the raster. They get worse as you turn the brightness control up, and disappear if you turn it down far enough. However, at that point you can barely make out the picture! High voltage, color, etc. okay. What is?—R. D., Cincinnati, OH.

Take the bottom plastic cover off the 3A3 socket, and push all of the socket pins firmly down while holding the tube's plate cap. In some cases, one of these will be just a little bit loose, making poor contact. This arcs when you increase the beam

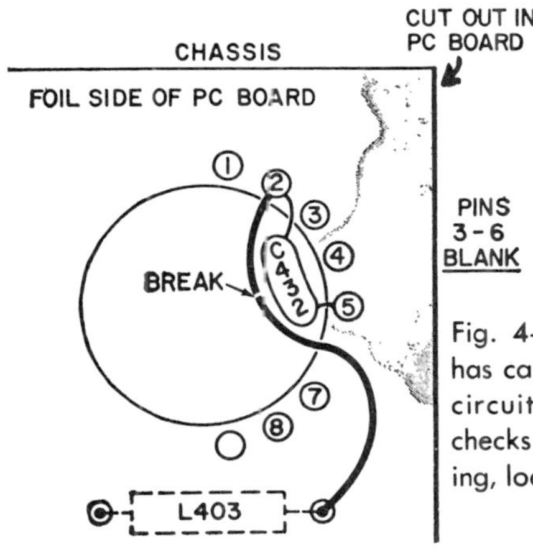

Fig. 4-16. A bent-over capacitor has caused more than a few short circuits. When everything else checks out and the HV is still missing, look for this.

current (turn up the brightness). The more current, the heavier the arcing.

Normally, this will hold until the tube is replaced or taken out of the socket again. However, the socket ought to be replaced the next time this set is in the shop.

Flashover In Focus Rectifier

The IV2 socket in an Admiral 3G11 color chassis went bad. I have replaced it, but I still get flashover from plate to filament. Doping it doesn't help. Could I replace it with A1X2, or a type with the plate on top?—V.V, Clifton, NJ.

Yes, but it might be a little tricky. How about using a solid-state focus rectifier? If you can find a good solid mounting, it will work nicely.

Also, you might check the RC network between terminals 9 and 8 (see Fig. 4-17) of the flyback that feeds the pulse to the focus rectifier plate. In this chassis, and some Philcos using a very similar circuit, arc-overs have been known to damage these resistors and capacitors. If so, this will cause flashovers later on. Be sure that the resistors are well insulated types, with at least a 3-watt rating.

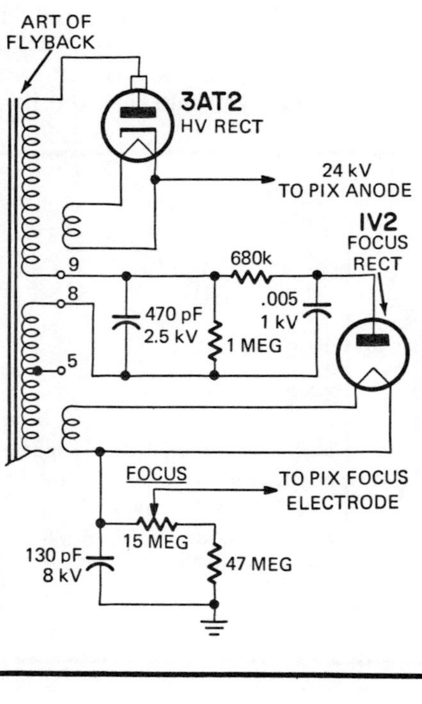

Fig. 4-17. Arcing can be caused by damaged components. For persistent problem, check RC network at pins 8 and 9 of flyback.

Bad Ultor Contact

I get horizontal streaks and flashes in this RCA CTC 17X. Video looks peculiar. The streaks are very thin and intermittent. The higher the brightness, the worse the streaking. I've checked all of the video tubes, voltages, etc., with no luck. This is a new one. Scope shows nothing unusual in the signal through the video stages.—K.N., Berkley, MI.

I know what you *might* have. The *internal connection* from the ultor button on the picture tube may be bad. The internal coating may have broken loose from the metal button.

If this happen, the high voltage arcs intermittently from button to coating, and this produces the thin horizontal streaks. Naturally, the higher the beam current (more brightness) the worse the arcing.

Clean the bell of the crt very well around the button. Fold the rubber cap back, hook up the lead, and turn out the room lights. Watch the ultor connector. If this has happened, you'll be able to see several tiny red arcs from the button to the coating, obviously *inside* the glass! Turn the brightness up and down while you're watching (or have someone else do it—this would take an awfully long arm). If the arcing *is* inside the tube and increases with an increase in brightness, that's it.

Only cure for this is a new picture tube, I'm afraid. Before you replace it, be sure to check all of the connections in the HV circuit; HV rectifier, socket, wiring, resistors, etc. A burned-out series resistor in the HV lead or an arced-over HV rectifier socket can imitate it very closely.

No Control Of Brightness

I can set up this RCA CTC 38XL to make a good picture, but the brightness control won't extinguish the raster. (I checked the brightness limiter, too.) The screen controls won't extinguish the beams, either. Everything looks normal; voltages pretty close. In SERVICE position, all I can get is a bright white line, but I can juggle the screen controls on a picture and come out pretty good. Something's still wrong, though.—T. S., Huntington, WV.

You'll find that you pix tube grids are too far *positive*. If the cathode voltages are close, and vary with the brightness controls, as you say, and the screen voltages are in the ballpark, there's only one thing left. The grid voltages, which are controlled by the color-difference amplifier stages, are so far positive that the cathode voltage cannot cut the beams off.

Needless to say, this will be something common to all three color-difference amplifiers. Check plate resistors, etc.

Chapter 5 Tuners, Noise, and AGC Problems

This may seem odd, in a technical book, but I'd like to go from *tech*nology to *psych*ology for a while. Don't worry; it has a real, practical application. It's about the use of *imagination* in our work. The best diagnosticians have a very vivid imagination, plus a thorough knowledge of the circuit they're working on. With *both* of these, they are able to make up a list of the many, many things that could be causing the symptoms they find. After this, all they have to do is check out the list, one item at a time, to find the cause.

Try it yourself, when you get a spare moment. Get a defective set; any kind. Also get a scratch pad. Open the set up, turn it on, and see what the main symptom is. No picture, weak rf, no sound, etc. Write this at the top of the sheet. Now, sit there and make up a list of everything you can think of that could possibly cause it. Number each item. You might start with the one you consider the most likely and go on from there.

Now start making tests. Check off each item as you go. If your imagination is active enough, you *will* have all of the possible causes on your list. However, I think you'll be suprised to find in many cases that there were, after all, one or more things you forgot to put down on your list.

I speak from a lot of experience. I have used this method for a long time; my shop is full of clipboards with notes on 'em. You may be a bit embarrassed the first few times. You'll find that you forgot to put down what turned out to be a very obvious cause, and this was the one. After a little practice with this method, I think you'll find your diagnostic skills sharpening up perceptibly. By backtracking, you will be able to see where you got off the correct line.

Above all else, this will help you to develop the most valuable attribute in all diagnostic work—*a completely open mind*. The cardinal sin, of course, is the "fixed diagnosis," made on the basis of no evidence whatsoever; just what you *think* it should be. If you commit this error, you'll find yourself testing not to find out what the real trouble is but *solely to verify your first opinion.* (For that's what this is—not a valid diagnosis, but just a personal opinion, until you have some hard facts and test results to back it up.)

For a good horrible example of this, you're sure that the loss of high voltage and the high current in the horizontal output tube is due to a shorted flyback. So you change the flyback; now you still have high current and no high voltage. Checking further, you find that the focus transformer was shorted, causing the whole trouble.

To use your imagination properly you'll have to reverse the tradition of the U.S. legal system. Everything must be considered guilty until proven innocent, by test. No matter how unlikely a certain part may seem, suspect it until you've proven definitely that it is good. When you get this result, check that one off your list and go on to the next one.

For another example, no high voltage, cathode current very low, screen voltage on output tube very low. Tube good, by replacement. Open flyback? No, by ohmmeter test. B+ supply at bottom of winding normal, so the damper is conducting. Imagination says, "This tube is acting as if it had no plate voltage, but it's there on the flyback." Final result: check plate lead; insulated cap, looking good; wire not damaged. Look *inside* the plate cap, and find that the wire has corroded and opened up. No plate voltage.

During WW II, I was servicing a B25 bomber which was ferrying through our field. It was parked on a taxi strip about half a mile from the tower. I pulled the SCR-522 VHF radio, took it to the shop, changed a tube, tuned it up, and called the tower. Fine. Time, 12:00 midnight. At 3: a.m., after two trips back to the shop, and a perfect response, the radio still would not work in the aircraft.

My test instruments showed that the transmitter had ample rf output, good modulation, and everything else, but I could not get the tower. Mikes, antenna, transmission line, and every other part of the ship's wiring checked. Since I had already tried cussing, I did the only thing left. I sat down on the runway and cried.

Sitting there in the cold Texas moonlight, imagination finally came to the rescue. From where I sat, I could see the "ax-handle" vhf antenna on the bottom of the fuselage. Looking past it, I could see that the antenna, the right landing-gear strut, and the tower were perfectly lined up. I said, "Imagination, shut up! That strut *couldn't* be attenuating that much signal. It couldn't be!" Imagination says "Try it. What have you got to lose except more sleep?"

I released the tail tiedown, fastened the rope to my jeep, and dragged the tail of the aircraft about 5 feet to one side. Crawling wearily back up into the cockpit, I picked up the mike. "Rosy Tower, this is Army 359. Radio check, please." Back came a roar: "Army 359, 5 by 5! Loud and clear!" "Thank you, Rosy Tower! Army 359 over and out! Goodnight!"

The moral of this is, of course, that after you've exhausted all of the reasonable, normal alternatives, sit back and let your imagination take it from there. There must be something you haven't thought of *yet*. Wind it up and let it fly, and sooner or later you'll get it. If all other ways have been exhausted, try a vivid imagination. Remember, no idea is too far-fetched to be worth investigating if all else has failed.

So now let's free our minds from snap judgments and conventional restrictions, and examine a few problems.

MIXER NOISE

One fairly minor problem, but one which customers often complain about, is "busy background." Translated, this means a sort of rippling, shimmering, or "working" in large smooth areas of the picture. It's noticeable in monochrome pictures and can produce funny effects in color pictures.

Very probably, this is due to *mixer noise* in the tuner! All mixers, being nonlinear things, generate a certain amount of unavoidable noise. The only way to get rid of it is to override it with a lot of rf signal—improve the signal-to-noise ratio, in other words. This is true of tube tuners and can happen in transistor tuners, too. In fact, in many hybrid or solid-state TV sets with transistor i-f stages, you'll get a very similar effect at some position of the agc.

The basic cure is even simpler. If the rf amplifier stage doesn't have enough gain, the signal-to-noise ratio will suffer. Outside of the normal reasons for this, such as weak tubes, bad resistors, etc., one often unsuspected case of trouble is the rf agc voltage.

Many sets use rf agc *delay* circuits. The idea of these is to hold the rf amplifier at its *maximum* gain until we have used up all of the "control action" we can in the i-f agc. At some point, we'll find that we simply can't reduce the gain of the i-f stages any further by applying more reverse bias (bias which will reduce the gain—either polarity in solid-state, negative in tubes). If we try, we get clipping and a few other unpleasant things.

So, when this point is reached, the rf agc delay should let go and allow agc voltage to be applied to the rf amplifier to reduce its gain. There are several ways of doing this. Some transistor agc circuits may have a special rf agc transistor, for delay purposes only. Of course, when this happens, the input rf signal is pretty strong.

This kind of "big signal" is normally found only in primary areas, close to the TV transmitter. However, in CATV and MATV systems, you'll often find signal levels as high as 1000 μV per channel. If the rf agc delay circuitry isn't working properly, away we go again.

AGC

After checking the common problems, such as weak tubes, if the trouble still shows up, check the agc circuit. One common and often unsuspected of agc problems is the big delay resistor (Fig. 5-1). This is used in many sets, and has been for a long time. Its purpose is to feed a small positive dc voltage to the agc bus, so the agc voltage has to go higher than a preset level before it can affect rf amplifier gain.

You'll find 20 to 30M resistances used here. Despite their large resistance value, these things are *critical*! They seem to have a tendency to drift and it takes only a very small percentage of drift to cause trouble. Because of the numerous shunt paths, you can't get a reliable check on them in-circuit. Lift one end and check them closely. For replacements, use 5% resistors, just for luck.

In transistor agc circuitry, a leaky gate or keyer transistor can upset things. (In one classic case, a leaky agc keyer transistor, reading about 12 μA leakage, managed to upset not only the agc, but the vertical oscillator! It was fed from the same regulated 12V source.)

Regardless of where the keying pulse is applied—base, emitter, or collector—a very minute leakage in the transistor will upset the bias and, of course, the agc.

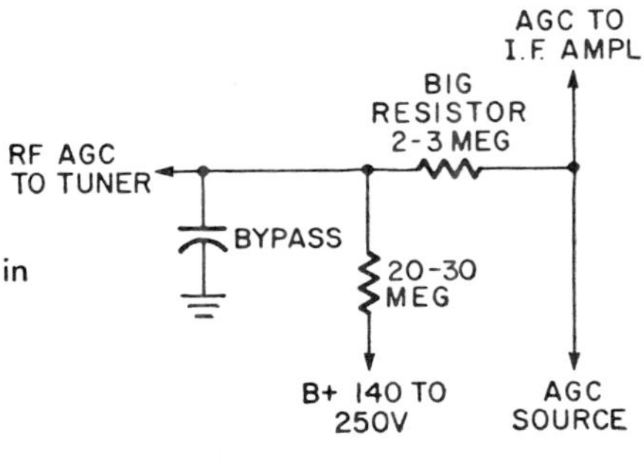

Fig. 5-1. Delay resistor in agc circuit.

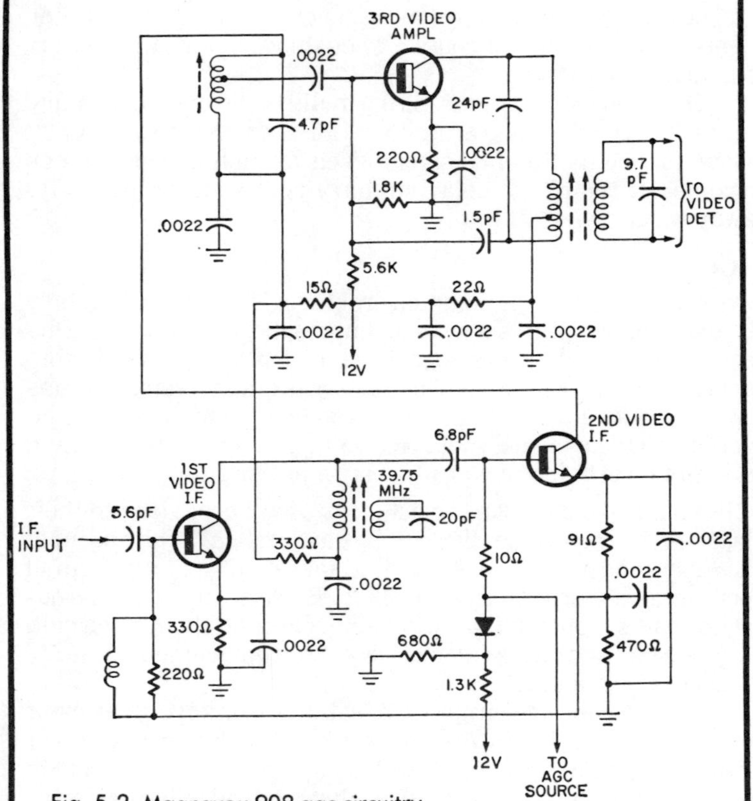

Fig. 5-2. Magnavox 908 agc circuitry.

Incidentally, you *can* clamp or override agc in transistor TV sets just as easily as you can in tube circuitry. Figure 5-2 shows the simplified i-f agc circuitry of a typical solid-state chassis (Magnavox 908). To override the agc, just connect your bias box to the i-f agc source. Set the bias box at the *no-signal* agc voltage and polarity (in this case about +3.0V). The no-signal voltage here is the agc voltage present when the controlled stages are in maximum-gain condition. Clamp it at that level and that's all you need. The main purpose of this test is to see whether the i-f can be made to pass a picture signal. Once you get this, you can vary the agc voltage a little to see how it works.

Finally, if you have checked everything and it still doesn't work, look for an open agc bypass, especially on the rf agc line.

The Open AGC Bypass

At regular intervals I embarrass myself by not doing what I keep telling others to do! There is one redeeming feature, though. I get to write up my mistakes and sell 'em, much to the envy of some friends.

The victim was an RCA CTC 38 color set. It has some dandy symptoms (all intermittent, of course): Horizontal pull, loss of sync, color smear, and a few more. It was brought in by a friend who had been working on it for a couple of days. He claimed that it acted up at home, but sat on his bench and played perfectly. (Normal, huh?) So, we set it up, hooked it to the test jig, and started in.

He was right; it did play very well. Voltage checks showed nothing unusual; the agc tube plate showed a normal reaction with and without signal. However, the agc adjustment showed a rather odd effect. Instead of going smoothly from whiteout to blackout, it had a sort of "hump in the middle" kind of action. Agc voltages seemed to be very close to spec. I tried to override the agc voltage with my bias box, without success. Checking the schematic showed why: my old box went only to 20V, and this one needed considerably more than that. So: "Oh, well. We'll do without it." (Mistake No. 1.)

At about this point in the proceedings, I accidentally found that the thing was *mechanically* intermittent. You could push on the bottom of the PC board, at certain places, and make the symptoms appear and disappear. This was going to be easy! Cold solder joint. After about 15 to 20 minutes of pushing, tapping, bending, and hammering on it, I decided that it was not going to be all that easy. Everything I *touched* made it cut in and out!

A close visual examination of the bottom of the PC board showed no obvious bad solder joints. So, I got down the shotgun; I resoldered all of the joints in the area that seemed to be the most sensitive. This helped just as much as anything I'd done so far (not at all!).

Quite a few fruitless voltage measurements and hammer tests later, I finally did what I should have, long before this. I got out the scope. Nothing in the way of excess ripple on the dc voltage supply lines; symptoms in or out. Keying pulse on agc tube plate clean, plenty of amplitude. So, there went that one. Signal on agc tube grid—oops! Here's something that isn't right! All bent up and badly distorted.

Back to the video detector output; signal there okay when board was pushed down, very badly distorted when board pushed up to make the symptom show up. Aha! This trouble is **not** in the sync, etc., where I'd been looking for it. It's farther back, in the i-f or tuner. (At this point, I am now beginning to show signs of having a normal mentality.)

Changing to the crystal-detector probe, I followed the signal back through the i-f, to the 1st i-f input. **Hmm**. Signal clean at that point, distorted when board pushed. Tuner trouble? No. Oh, oh. What *controls* the i-f? The agc. So, how about the agc line? (Right at this point, I'm looking pretty good. About time, too!).

Going back to the low-capacitance probe, I scoped the agc line. Oh, wow! With the symptoms out, clean. With the symptom in, the dangedest mess of stuff you ever saw. The scope photo (Fig. 5-3) shows what this pattern looked like. This is just a *little* out of the ordinary! The agc line should have nothing but pure dc on it. Not this mess of horizontal and vertical pulses, at about 5.0V p-p.

After I stopped hitting myself on the head with the rubber hammer (which I keep on the bench for this purpose), the real nature of the trouble dawned on me. There are all sorts of pulses and things on the agc line, at it's "beginning." What parts are used to take them out? Bypass capacitors. (Boy! How complicated can you get?) What would happen if one of these bypass capacitors were to **open**? I'd have exactly the same kind of mess on the agc line that I had just seen. Ergo, the intermittent that I had been fruitlessly chasing for so long was now identified. It was one of the three agc bypass capacitors. Had to be.

Checking these capacitors carefully showed that a 0.001 μF capacitor directly connected to the emitter of the agc transistor had a very bad ground "connection." This joint just happened to be under a couple of wires on the bottom of the PC board, and I hadn't seen it. When it opened, being right on the agc input to the i-f, it allowed all of those pulses to be applied to the i-f bases. There, they caused the mother and father of all

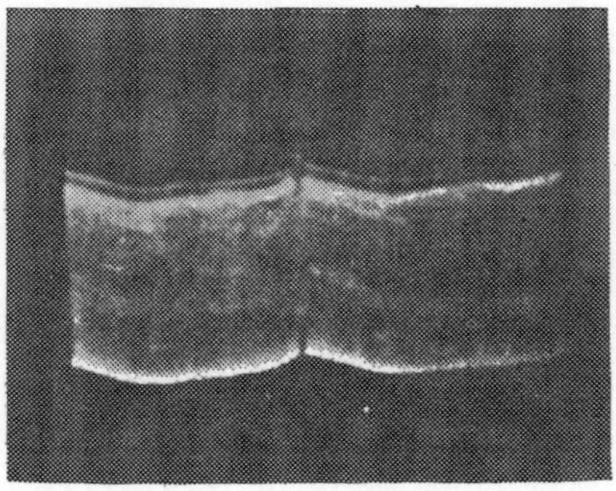

Fig. 5-3. This is what the brain found on the agc line, when it went looking for it! 5.0V p-p of assorted mess, on the i-f agc!

feedbacks, and I got all of those different symptoms at the same time.

The moral of this is very simple. When I stopped using my muscles—pushing and hammering on the board, and began using my head—to make a logical *analysis* of the symptoms, I could trace the fault back through the circuits to the point where it *began*. Then, everything cleared up in a hurry! Total elapsed time, using second method, not more than 10 minutes; time wasted using first method, about two hours.

So, **stop** and **think**. You **know** how these things work, and "what causes certain reactions." If you'll make a few logical tests, and then interpret the results correctly, you'll get along a lot better, and be embarrassed much less frequently!

Recognizing AGC Norms

The stock agc rectifies the tips of the horizontal sync pulses, develops a dc voltage proportional to the amplitude, then feeds this to the i-f grids. From the feedpoint, through some kind of delay circuit, the same voltage is fed to the rf stage, after the i-f agc has gone as far as it can to control signal level.

We see some interesting things in transistor TV sets. The agc path in Fig. 5-4 turns somersaults on the way from the agc keyer to the i-f/rf. This happens because we can have either positive-going or negative-going agc in transistor circuits! It can be forward agc where the transistor is biased so that forward bias *reduces* the gain, or reverse agc, which is the type used with tubes. The bias developed is in a direction which tends to cut off the tube or transistor.

The diagram shows a compound agc circuit, as used in Admiral 1H5 and 2H5 transistor TVs. Looks like quite a bag of worms! However, there's no need for alarm. Take it one piece at a time, and the circuit is just as easy to check as any of them. There is nothing in here but a bunch of perfectly *normal* reactions! For example, if you increase the forward bias on a transistor base (if it's forward agc) you have less gain and so on. In any case, you'll be able to see the reaction—there will be a distinct *change* in the agc voltage when the signal is applied or taken off. Follow these reactions through the circuit and you'll soon pin down any trouble. Find the point where you *do not see* normal reactions and there you are.

Let's go through this one step at a time. You'll find this, and similar circuits, in a lot of chassis, of different makes. The diagram shows only the agc circuit and the working parts. Some have been left out for clarity.

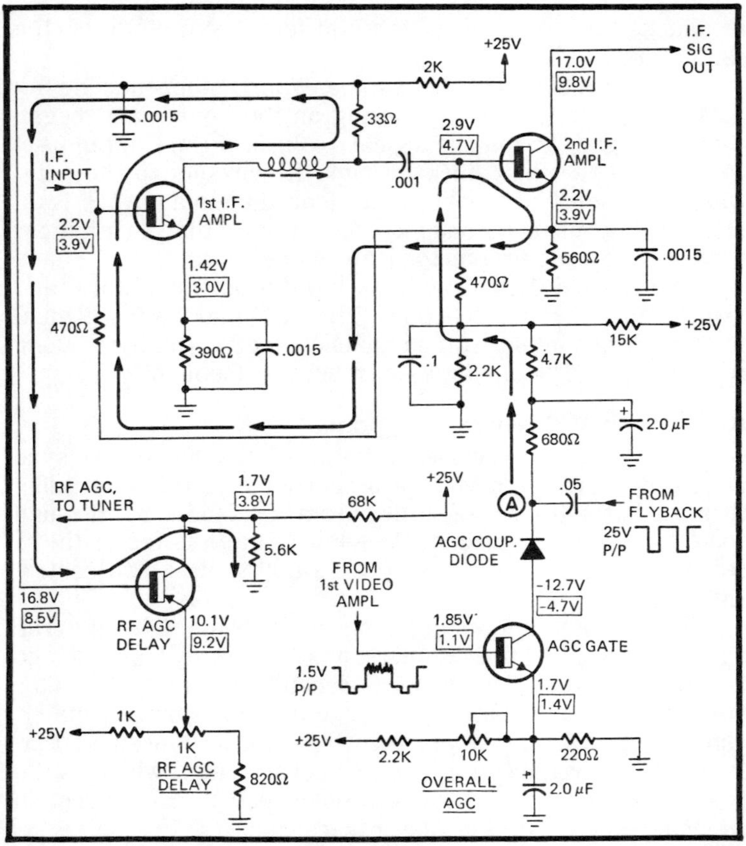

Fig. 5-4. Agc signal flow in typical transistor circuit. If you're accustomed to looking at tube circuits, this can seem almost frightening at first.

We start with a stock agc keyer, the *agc gate*. About 1.5V peak-to-peak video is applied to its base. A keying pulse from the flyback, shaped to a sawtooth, is fed to the collector through a gate diode. The conduction of the keyer transistor is adjusted by varying its emitter voltage. This is done by the overall agc control—part of a voltage-divider from the +25V source to ground. The agc voltage is picked off at the diode cathode, filtered by a 2 μF capacitor, and starts out on its long journey (point A in Fig. 5-4).

From here it goes through a voltage divider to the base of the second video i-f transistor. It is an npn, forward-agc controlled. The agc voltage at this point goes *more positive* with an increase in signal. So the higher the signal level, the greater the gain reduction in this stage. From the emitter of this stage, we pick up another varying voltage. It is developed by the emitter—collector current flowing through the emitter resistor. When the agc makes the base more forward biased, the current and the voltage goes more positive with a rise in signal.

Off we go again. This time to the base of the first i-f transistor. It is also an npn. Since the agc voltage is still positive-going, this stage is also forward-agc biased. We have now control of the first and second i-f stages, with only one connection to the agc itself. Next, we need some agc voltage for the rf stages. It must be delayed, so it will have no effect (leave the rf stage at maximum gain) until the i-f agc has done all it can to hold the signal level down. We do this with an extra rf-agc-delay transistor.

The i-f voltage causes a change in the emitter-collector current of the first i-f transistor, just as before. Now, though, we take off at a different point. We break the collector load resistor into two parts: 2000 ohms from the +25V supply, and 33 ohms. The rf-agc-delay transistor's base is connected to the junction of the two. When the first i-f transistor's collector current rises with increasing agc voltage, there is a voltage drop across both resistors. Most of the drop appears across the larger resistor. Now if the i-f voltage increases, the agc voltage goes *more negative*. The rf-agc-delay transistor is a pnp. So though we have changed polarity, we still have a *forward agc* voltage.

The +25V supply is fed to the rf-agc-delay transistor emitter, through still another voltage divider. The collector goes to ground, through another one, also returned to the +25V supply. So the voltage developed across the 5600-ohm part of the collector divider goes more positive (greater current through the resistor) with an increase in signal. It is the rf agc control voltage. It is fed directly to the base of the rf amplifier transistor, an npn. Once again, this is a forward-agc reaction.

A separate control is provided from the voltage divider in the rf-agc-delay transistor's emitter circuit. This control can be set to get whatever delay we want. It holds off conduction of the rf-agc-delay transistor until the i-f agc voltage reaches a certain level. From this point on, the rf agc takes hold, to prevent overloading the i-f stages.

How do we service this circuit? It's a lot easier than it looks. The first thing we do is go directly to the source of the *initial reaction*, the agc keyer (gate) itself. Read the dc voltage at point A. Then hook up the antenna and see if it *varies*. (It should go more positive, of course.) If it doesn't, check all the parts and signals in this circuit before going any farther. Look for the keying pulse. It is a negative-going 25V peak-to-peak pulse. Check the dc voltages on the transistor and all resistors.

If the thing refuses to vary voltages when the agc control is turned, check the transistor for *leakage*. You'll have to take it out of the circuit to do this, of course. It doesn't take much leakage to upset things here. If you find any leakage, replace the transistor.

Once you get the correct reaction on the agc diode, follow it through the second i-f transistor. Use the same basic methods and tests. You *must* get the normal *voltage variation* on the emitter of this one. From here, onward and upward to the first i-f, where you do it all again. This time look for the voltage variation on the collector.

From here, we go to the base of the rf-agc-delay transistor. Look for voltage variation again. If you see it on the base, but nothing happens on the collector, check the transistor for leakage. There are also some parts in the tuner, such as another 2 μF capacitor, resistors, and the rf transistor, which can upset things. However, after what you've been through to get this far, this should be easy!

Dc voltages shown in Fig. 5-4 are for a no-signal condition. The voltages in the little boxes are typical voltages you should see with a fairly strong signal applied to the input. By the way, one of the quick-disconnect "clothespin" antenna clips is handy for making these tests. With the clip in one hand, and the VTVM probe in the other, away you go!

Finally, and a lot of fellows don't know this—you *can* clamp a transistor agc circuit with a *bias box* for testing, just as you do with tube sets. Check the no-signal voltage at any given point and adjust the bias box to that value. This is the "condition of maximum gain" for that particular point in the circuit. May help out in cases where you have doubts.

TV BOOSTERS

Noise problems don't always point to a tuner problem. There are a lot of TV *boosters* around; mostly transistor types by now, I suppose, although there are still some tube types in use. Most are used in fringe areas, but you'll even find them in suburban areas. At any rate, there are enough to make servicing them worthwhile, if we can find a fast and easy way to check them.

This we can do, using only two pieces of test equipment that should be found in all shops; the color-bar generator and a scope with a crystal-detector probe. You do *not* have to take

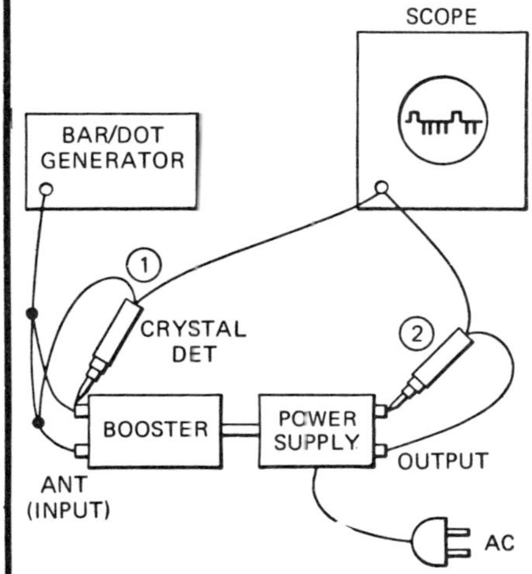

Fig. 5-5. Equipment setup for checking TV boosters: a bar/dot generator, scope, and crystal-diode probe.

the booster apart to check it. All you need is the booster and its power supply. Connect them together with about 3 feet of 300-ohm twinlead and plug it in.

Hook the rf output of the generator, set for a color-bar pattern, to the antenna or input terminals of the booster. Connect the scope detector probe across this, as in Fig. 5-5. Turn the generator rf gain control full on, and turn the scope's vertical gain to maximum. (You won't have much signal to work with here.) Set the scope's horizontal sweep to display several lines of video.

You may not get much deflection on this test; the signal is at a pretty low level. However, if you get even a slight "thickening" of the scope trace, fine. Turn the generator rf gain control up and down, and see. Now move the scope probe to the power supply output terminals, where the lead goes to the TV set. If the booster is working, you should now be able to see a fairly good-sized display. In an actual check of a small two-transistor unit, I got what looked like 0.1 cm at the input, and 0.8 cm at the output. Of course, you do not move either the generator's rf gain control or the scope's vertical gain control after getting the input reading.

From these readings, you can make a very accurate estimate of the amplifier's signal voltage gain. Remember this will be in decibels, and it will be voltage, since that's what we want to amplify. The reading on the scope can be in inches, centimeters, or even in arbitrary lines on the graticule; we

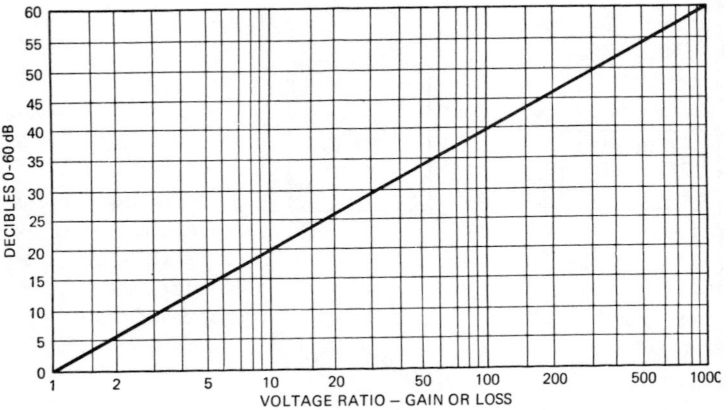

Fig. 5-6. How a function such as voltage ratio relates logarithmically to a linear decibel scale. Note that each tenfold increase of voltage is a linear increase of 20 dB. And doubling any given voltage results in a 6 dB improvement.

couldn't care less. All we want are two figures: the input and output.

The chart of Fig. 5-6 will help. In my test, I got about a 1:8 *ratio*, which comes out at about 18 dB (pretty good for a little rascal like this one). The same method will work with the higher-gain amplifiers, such as MATV line amplifiers, and so on. Your readings will be higher, but that's the only difference.

As I said, this is a rough test, but good enough. For example, if the specs on the booster showed a gain of either 15 or 20 dB, I would have been well inside the ballpark.

This test is also useful for trying out amplifiers after a bad transistor has been replaced. If you can get a gain anywhere near the original figures, fine. To check for bandpass, tune the bar/dot generator from one end of its coverage range to the other; this is assuming that you have a variable-tuned bar/dot generator such as Heathkit IG-28. Same check could be made with a tunable rf signal generator; use 400 Hz amplitude modulation, and tune it across the low and high TV bands. A marker generator is good for this.

If you have input but no output, the booster is completely dead. The first thing to check is the dc power supply. The typical booster power supply feeds on a low ac voltage up the lead-in (16–24V, etc.). At the amplifier, this is rectified and used for the dc supply. Transistor boosters use only a few milliamperes; tube types must provide for the heater voltage, and the current will be greater. Figure 5-7 shows a typical supply. The TV signal is fed into and taken off the lead-in through rf chokes, little coupling capacitors, etc.

Check the ac output of the power supply first. If the booster was hit by lightning, the fuse may be blown. Some of these don't have fuses, although all of them *should*. If the ac voltage is okay, check the dc voltage on the output of the diode in the amplifier chassis. If it is okay, check dc voltages on the transistors. This will show up shorted transistors, open transistors, etc. The ohmmeter will help here, too.

If a transistor has been replaced, and the unit seem to be working normally, make one final check before declaring it fixed. Hook up a TV set, and connect the booster output to its antenna terminals. Now, hold the end of your bench antenna lead-in close to the input terminals of the booster.

This simulates a very weak signal. The main thing you want to watch for here is the presence of hum bars in the signal. If one of the tiny electrolytic filter capacitors in the booster power supply is open, you can get the mother and father of all hum bars in the picture. It's much easier to fix now than it would be after you've climbed the mast and put it back.

Be sure to check the transmission line at the home if the booster shows any sign of lightning damage. If there has been

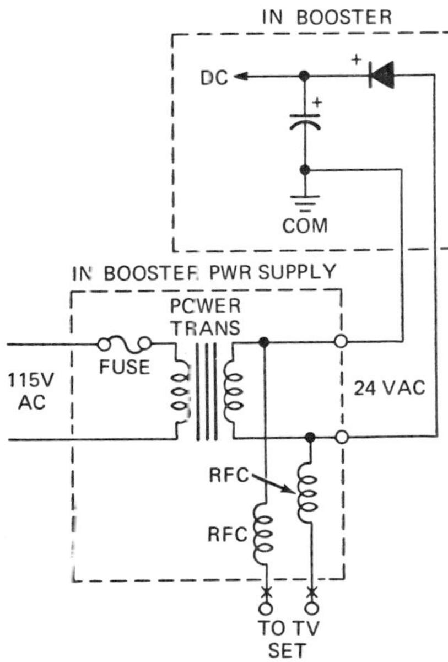

Fig. 5-7. Insert 0.01 μF capacitors at X-X.

a close hit, some of the conductors in the twinlead may have been burned in two, or coax may have been melted and shorted. If the antenna has taken a direct hit, there will be no doubt about it; the booster will probably be blown to bits internally, and completely irreparable.

INTERFERENCE

So as not to overlook noise sources other than those generated internally, let's turn our attention to interface. Outside of certain commercials, one of the things that annoys TV viewers the most is interference or noise in the picture. It comes in all shapes and sizes; from older cars, home electrical appliances, and so on. Anything in the house can be tracked down and identified by simply turning it on and off. Filtering can then be used to clear it up. Some interference can come from outside sources, though. We used to call this *hash*, but now I suppose we should say noise pollution. Either one is correct.

An irate reader wrote me: "*I've got so much interference on my TV set that I don't like to watch it. There are two wide bands of dots and streaks that roll slowly up the picture. They interfere with the sound and even make the picture roll at times. Looks like ignition noise, but I don't think it is. What can I do about it?*"

Complain loudly to the power company. I've seen this a great many times. The polite name for it is hardware noise. It's caused by old or defective equipment on the ac lines feeding your house. You'll notice that the interference shows two bands of dots, meaning that it has a characteristic 120 Hz frequency. Without going into too much detail, it's typically due to a *corona* discharge. Not on your home's service wiring, which is 320V maximum in most places, but on the feeders or tertiary lines which supply the pole stepdown transformers. These lines run around 7600V, and often even higher.

Several things can cause it, including defective or cracked insulators; but probably the most frequent offender is the terminals on the transformers themselves. If these are old or not correctly designed, the corona discharge takes place *inside* the transformer, above the oil, from the shank of the bolt to the metal case. I have seen some horrible examples of these after they'd been replaced, and the shank of the terminal bolt is actually pitted and eaten away by the arcing. There are newer insulator types, designed to prevent this; they have plastic insulation, etc. on the shank of the bolt. Your power company probably has a training film on this, which shows all of the effects and how to cure them.

The reason for the 120 Hz effect is simple. The corona discharge takes place twice during each cycle of the voltage; once on the positive peak and again on the negative peak. Each peak generates the corona discharge, which in turn generates a burst of high-amplitude rf "signals," which cover all frequencies. We're going to do something with this effect in a moment.

The first step should be to file a complaint, in writing, with your local power company. Tell them you have hardware noise, and they'll know what you're talking about. Power companies have noise locating equipment, and it shouldn't take them long to find the offending unit.

If you want to try finding it yourself, there are a couple of ways. Drive around the area with a radio equipped car, or even a portable radio. Set the dial off-station, somewhere near 800 kHz. For some reason, this noise has a minor peak at that frequency (you'll final major peaks in the VHF band, particularly around 50 MHz). When you find a place where the noise is very loud, turn the dial of the set back and forth. It will usually be of such a high amplitude that you won't be able to use "loudness" alone to pin it down.

When you get closer to the source, the noise will start to spread; it will cover more and more of the dial. For example, if you get noise from 600 to 900 kHz, you're getting closer, but you're not there yet. Keep on looking and tuning, and when you can get nothing but noise from one end of the dial to the other, you're getting warm—really warm.

Find the nearest pole. If it has a transformer on it, so much the better. Listen to the noise, and hit the pole a sound whack with heavy hammer. If this changes the noise, there you are.

Second method: If you have a highly directional TV antenna with a rotator, turn it all the way around and watch the screen (channel 2 will probably be most effective, unless there is a strong local station at that frequency). Find the direction where the noise is most intense. Make a line along this bearing, on a city map. Now locate a friend with an antenna of the same type. Take a bearing on the noise with his antenna. Draw this on the map. At the point where the lines cross, there's the noise. If you can get a third bearing, preferably at a point distant from the first two, you can pin it down even more closely. I used this method on a very bad case some time ago and pinned it down to a very bad insulator on a pole almost two miles away! Some nut had shot the insulator with a high-powered rifle!

Because power lines make excellent long-wire antennas, you may run into some ambiguity in your bearings. In fact, it's more likely that you'll see a triangle instead of a single point where they cross. At any rate, you're in the ballpark.

Go to the general area where the noise is originating and then switch to the car radio or portable. Use the frequency-spread of the noise as the indicator and follow the power line until you find the pole where the noise covers the whole dial. There you are. Look for a little metal tab or label on the pole. This will give you the number, and you can report it to the power company.

READER QUESTIONS

Low Video, Bad Smear

The picture comes on very slowly in an old Motorola TS-544. The raster lights up in the normal time, but it takes almost a minute for the picture to show up. When it does, it is very badly smeared horizontally. Is this an agc problem?—J.G., Mena, AR.

I doubt it very much; not with these symptoms. Read the dc voltages on the plate and screen of the 12BY7 video amplifier. You'll probably find the plate voltage pretty low, say down around 75 to 80 volts (normal: +175V).

If this is the case, check all four peaking coils in the plate circuit, and that 5600-ohm resistor. (See Fig. 5-8.) If any of these coils reads more than a very small resistance, the coil is open and you're reading the shunt resistor. Replace these. The 5600-ohm resistor has been known to go way up in value, also; it causes almost the same effect.

Snowy I-f

In a Magnavox transistor TV, I've run into quite a few that develop excessive snow. This seems to happen within a few weeks after they have been sold. Do you have any ideas as to what could be done to help this?—A.C., MS.

For one thing, transistor i-f's in many sets will have one position of the agc that will cause snow, even on a strong signal. You might try readjusting the agc in the home, to see if this helps. Alternate: Cook these sets as long as possible while they're still on the floor.

I have heard one, recently, which I pass along without comment. A field engineer for a well-known test-instrument company says that transistors will *change* in characteristics during the first few hours of use. Your experience might tend

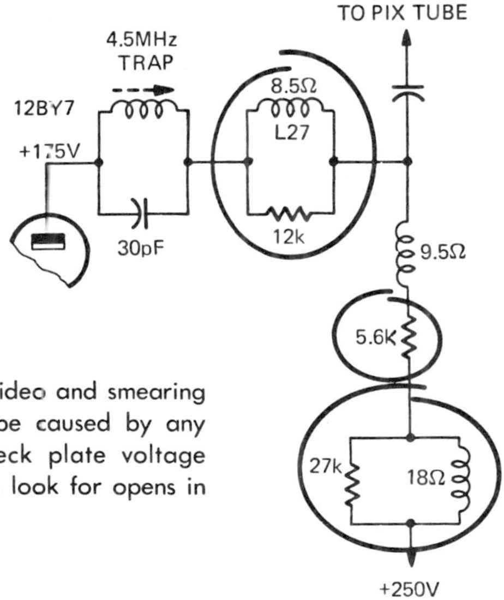

Fig. 5-8. Low-video and smearing problems can be caused by any open coil. Check plate voltage first; if it's low, look for opens in coils.

to bear this out; it would take some careful experimentation to verify it.

Agc Problems

One set came in with a bad flyback. I replaced it. Now I have a nice clean raster and plenty of high voltage, but no picture! Just a clear raster. I am reading −80V on the plate of the agc tube. The terminal markings were slightly different on the replacement flyback, but I followed instructions in replacing it, and that much seems to be okay. Why do I get this high negative voltage on my agc tube plate?—J.D., San Angelo, TX.

I have a sneaking feeling that I *know* why. Check your *keying pulse* on the agc tube plate. You'll probably find that it isn't there at all, or that it's of the wrong polarity. This often happens where there have been terminal changes on replacement flybacks, or in a few cases where someone has picked up the *wrong* flyback transformer in the parts depot.

Band Of Ripples

I've got some of the most weird symptoms I've ever seen in this Motorola Quasar II TS-929 chassis. With a picture, there's a vertical "band of ripples" right in the middle of the screen, flickering with all colors! If I take the antenna off, I get very thick black vertical bars, odd shaped, like Barkhausen, but definitely not; they're all over the screen! I scoped the horizontal sync, and the sync pulse goes all to pieces when the antenna is disconnected.—R.T., Hatfield, AR.

I've seen this before. Uncommon, but it happens. Look for a very small rf choke, encapsulated, mounted on the plate cap of the 6LF6 horizontal output tube. Take a piece of wire and short this choke! This will clear it up.

Apparently, this choke causes the horizontal output stage to ring heavily at some high frequency, and generate parasitic oscillations. Shunt it, and then quit! (Usually, the rf choke in the horizontal output plate lead suppresses spurious oscillations. However, in this case, it appears that the choke can cause HF or VHF oscillations.)

Tuner Trouble

I can get an i-f signal through from the input, and see good clear bars on the screen. No stations from the antenna. I suspect the tuner. Right?—R. B. M., Miami, FL.

Right! Most likely suspect: one of the transistors in the tuner. Check for correct dc voltages, and especially for correct or nearly correct *emitter* voltages. If you find one with collector and base voltages close, but no emitter voltage, that's the one: transistor probably open.

The tuner on this Zenith 16Z7C50Z has one of the funniest problems I've ever seen. I get channel 7 on the 6 setting, and so on all up the line. Fine tuning seems to be okay. You can't get channel 7 on 6, or can you?—T. S., Wickes, AR.

Nope. The cure for this is simple. Look at the back of the tuner. The *indicator* is driven by a small plastic pinion gear, spring-loaded, and a large plastic gear on the tuner shaft. This little gear can *slip* and skip a few teeth, and then your dial says channel 6 when the tuner is really on 7! Just lift the little gear against the spring, move the indicator to the right number, and let go.

It *is* possible to fine-tune another channel, especially on the high bands; 9 on 10, etc. Most fine tuners have this much range. However, if the dial is one number off all the way around, look for mechanical trouble.

AGC Too High

When I turn this Sears 9168 color set on, the picture comes on for a second, and then it whites out. I read a very high negative voltage (−60V) on the 12A V6 agc tube plate. Where's this coming from?—G.S., Highland. CA.

You're in the right place! This very high agc voltage is cutting of the i-f. Check to see that you do not have the proper keying pulse on the plate of this tube. Should be 200V p-p. Also, check the 33-megohm resistor from agc to +125V. Could be open.

However, I believe I'd be more apt to suspect an open agc bypass capacitor somewhere in there. This would let the agc go into oscillation, and build up that excessive negative voltage on the plate. Check the cathode voltages, through that voltage divider and the agc control. The 47,000-ohm resistor to ground may be open or increased in value.

RF AGC Trouble

This Motorola TS-454YC portable has developed a peculiar condition. Acts very much like agc trouble, but there is no agc adjustment on this chassis. All voltages seem to be okay and tube replacement didn't help.—D.L. Wash. D.C.

Yep. This could be a thermal trouble in a printed-circuit unit in the agc circuit (Fig. 5-9). The 12.0-megohm resistor could be drifting from heat, upsetting the rf agc bias. It can be replaced with a duplicate, or with separate series resistors if you can't locate the right value. Thermal drift in any of these resistors would upset agc.

No Horizontal Sync

I've an old Zenith, a 16D21Q, with no horizontal sync. The video waveform on the output looks funny, and the dc voltages around the 6BU8; no help—P.Q.R., Pine Bluff, AR.

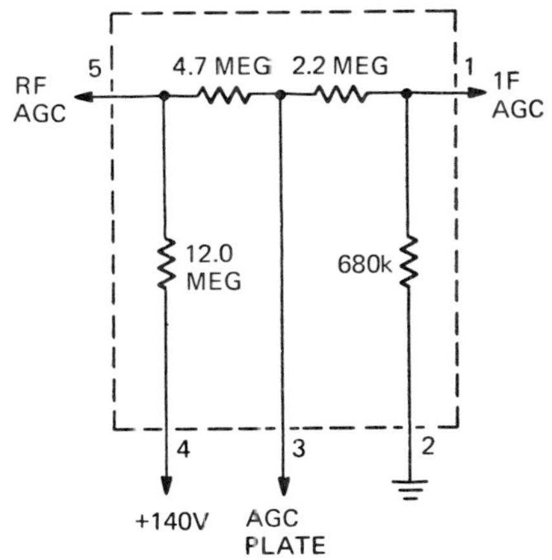

Fig. 5-9. The high-value series resistor in the B+ line can drift; when it does, agc problems will be many (and probably varied).

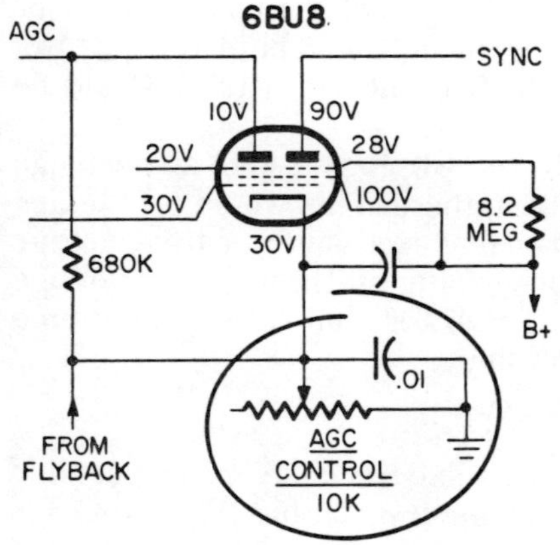

Fig. 5-10. When dc voltages on the tube look peculiar, start following the cathode circuit to ground. The resistance should be fairly low from cathode to ground—in this case no more than a few thousand ohms.

Check continuity on the agc tube! If all the voltages on this tube are high, something is wrong in the bias, and the only thing in the cathode circuit is that control! The diagram of Fig. 5-10 shows it. If this is open, so that your cathode voltage is too high, this will upset not only sync, but agc.

Screwball Reactions, TCA KCS-136

I've an RCA KCS-136 chassis that's got some of the screwiest reactions I've ever seen! It's got a pale picture, lousy sync, and the agc voltage is much too high, negative. This is not bad; but when I pulled the 6GX6 audio detector tube

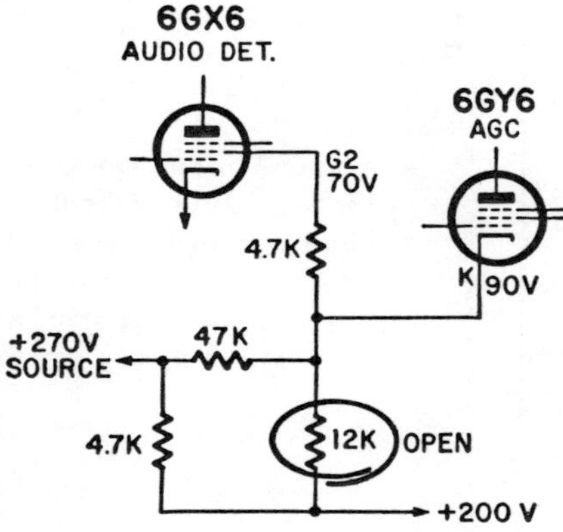

Fig. 5-11. In the RCA KCS-136 chassis, there's a 12K resistor used by part of a voltage divider. When it opens, the tube's screen grid offers an excessive load.

(I forget why, but I did), the picture got better! Cleared up a lot! Riddle me that one!—R. F., Boston, MA.

Well the crystal ball nearly popped its circuit breaker on this one, but I found it in the file! It happened to me some time ago, with almost all the same reactions.

Your major trouble *is* agc, as you thought. So far, so good. The "picture gets better when the sound tube is pulled" symptom is due to the simple (after you find it!) fact that the agc tube is supplied from a little voltage divider between the +200V and +270V lines. (See Fig. 5-11.) *So is the 6GX6 audio detector!* Taking the screen-grid load off the voltage divider changed the voltage. You'll find that the 12K resistor in the divider is open, I believe. This changes the agc tube cathode voltage and of course upsets the dc voltages around that whole stage. Isn't it remarkable how simple these are *afterward*?

Chapter 6 Diagnosing Trouble & Shaking Down Intermittents

Diagnosis can be simplified and made more accurate if you pay attention to procedure—put "first things first."

This is really simple. The first thing to do is "state the problem." There are two steps to this—*examination* and *identification*. In other words, turn the set on and see just what it is doing or not doing. From the results of this, we get our first bit of data. This tells us what section of the set to look into. When we find that we have a certain problem, as shown by the symptoms, then we go to the final step. This is the *elimination* step, and it's the longest one.

As we all know too well, there is no such thing as a problem in electronic equipment with only one possible cause! There may be one somewhere, but I haven't found it. So, the answer to our question "What's causing...?" is *always multiple choice*! We must check out all possible answers.

Take the first two steps: identify the nature and probable location of the fault. Then make out a mental list of the circuits, parts, etc. that could cause it. Now we patiently, calmly, and impartially test each one of these. Somewhere along the line, we'll find the one we're looking for. Notice the last word in there—"impartially."

That's the key word! The most valuable asset we can have during diagnosis is a completely *open mind*. We know that there is a faulty part in there somewhere—what difference does it make which one it is? None at all.

It's surprising how contrary the human mind can be. In far too many instances, we make up our mind that a certain thing is causing the trouble. Then we make tests, not to locate the real cause, but to verify our first guess! (And at this point, without much data, that's all it is—a guess!).

We have become *partial* to a certain answer, not impartial. This ruins our scientific objective approach to the whole problem. So when you run into a problem, the first thing to do is "cool it." Sit there and look at it, and note what it is actually doing.

From this point, you make up that list of possible causes. Then you slowly, carefully, and methodically check them until you find the bad part. This list *must* be based on a thorough knowledge of how the circuit works. If you do not have this, let the thing alone until you can dig out a book and read up on it. Sitting there and "poking at it" with a voltmeter will get you nowhere fast!

I've seen this happen. In fact, I do it myself far too often. When I work out a job, I make detailed notes of every test made, the reason for it, and the results. When I run into troubles, I can see where I went wrong. In every case, I find that somewhere in there I "froze" on a mistaken diagnosis, and then started looking for evidence to "make me be right,' instead of really trying to find out what was wrong.

You can try it yourself. Get a clipboard and some paper, and start writing down each step you take, and the reasons, and the results. Even if it does nothing but keep you from repeating yourself, it's worthwhile. Don't overlook the simple things, either. (I worked on a very complicated problem for at least an hour, only to find out that the reason I had no signal output from the stage was due to zero screen-grid voltage on the tube!)

There's one more common problem; the complaint that "I've checked everything in that circuit and it still doesn't work!" There are two possible answers: (1) you *haven't checked everything*; and (2) the trouble isn't in that circuit at all! Go and look for things which could *affect* this circuit. Example: all dc voltages, the tubes, capacitors, transformers are okay in a bandpass amplifier, but you can't get a color signal through it. The trouble turns out to be a high negative voltage on the grid, cutting the tube off; the color killer is the section at fault!

Other problems can have much more complex answers, but this just means that our list of "possibles" is that much longer. For a good example, there are a great many things that could cause the cathode current of a horizontal output tube in a color set to run far above normal. All you have to do is make your little list and patiently eliminate them one at a time, and you'll find it. Just for fun, take a piece of paper and make out this list. Then look at it and see how many actual cases you can remember where something *you have left out* was causing it! When you get through, pin the list up over your bench.

THIN HORIZONTAL LINE

This one happened to me, and it illustrates how reasoning intelligence (and a good deal of luck) will help. There was nothing but a very thin horizontal line on the screen of this Zenith portable; only one scanning line thick, really.

Applying the standard tests, I pulled the yoke and read the resistance; fine. Resistance of yoke section of vertical output transformer also fine: 9 ohms, just as in the book. Tube scope, no signal of any kind at any point in oscillator circuit.

Check all capacitors and resistors; they're good too. Now, try some of the clever little tests you keep telling other people to do, Darr. Clip lead to heater voltage at a 6V point. Feed this to grid of 10GK6 output section. Hmmm. Nothing. Still very thin line. Try higher voltage: 35V ac on grid; still nothing on screen. Up scope. Signal reads 35V (rms) on grid. On plate, about 2V peak-to-peak, at the most.

Well! Recheck dc voltages on output section just for luck. All present and accounted for. Now what's going on? Sit back and try something brand new—*THINK! What* could cause an amplifier tube, with plenty of signal on its grid, to have absolutely zero gain? One thing—a *load impedance* of absolutely zero ohms.

Pick up ohmmeter, turn set off, read resistance of primary section of vertical output transformer. Is it 275 ohms, as it says on the schematic? It is not! It is just what you thought; absolutely zero. Replace transformer, everything lovely. (This was apparently a lead short, from the very low resistance. It was in warranty, so I replaced it.)

There *is* always a legitimate, theoretically correct cause for everything, if you can just find it.

ASKING THE RIGHT QUESTIONS

Diagnosis is comparatively easy if you remember the rules: What is happening? What can we do to make it stop happening? The last is the "fix" that makes the thing work again.

The set is a Sylvania DO-3 color TV. The visible symptoms are: no vertical sweep, bright line across the screen. If I throw the *service* switch, about $^2/_3$ of the line disappears. Plenty of brightness. In fact, I had to turn the screens down so that I wouldn't have too bright a line while I was working. There is a very slight burnt-resistor odor and a low hum in the sound, with an occasional motorboating.

Taking the back off, I see a 1200-ohm resistor very badly scorched. It is in the cathode of the 21'lu8 vertical output section. Resistance measurement shows about 200 ohms to ground, but there is a 1500-ohm resistor and a convergence winding in parallel. So we clip one end loose and find that it reads about 500 ohms now.

I tack another resistor in here, for luck, and go on. Now, measuring the voltage on the vertical output tube plate, pin 4, I get about 15V. This is bad, alone, but in this case, worse, for this is 15V *negative*! Screen grid the same. What's going on here?

Diagnosis 1

At this point, I made some standard checks, just for luck; and substituted the 21LU8 tube, although it tested okay. No help.

From experience, this "negative plate voltage on vertical output" symptom meant one thing; the whole agc stage was in violent oscillation. The net result of this is to develop a voltage on that plate that reads negative to ground. Support for this was in the sound: the low hum and motorboating.

Although I was pretty sure by this time, I ran a couple of confirming tests. Checking on the 21LU8 plate with a scope, I found a high amplitude waveform like the one shown in Fig. 6-1. It was about 100V peak-to-peak, and turned out to be sawtooth waveforms at about 15.750 kHz. Next, I grabbed the electrolytic capacitor substitution box, and hung a big capacitor on the B+ test point that feeds the output stage (**L** on the schematic). Same waveform here.

When I pushed the button, the hum dropped and the raster filled up. However, this didn't clear up the hum completely, and there was a very bad rippling on the raster. Now, you have all of the information necessary for a complete diagnosis.

Diagnosis 2

There is one thing that causes most of the "violent oscillation" troubles; a bad filter capacitor. This is what I

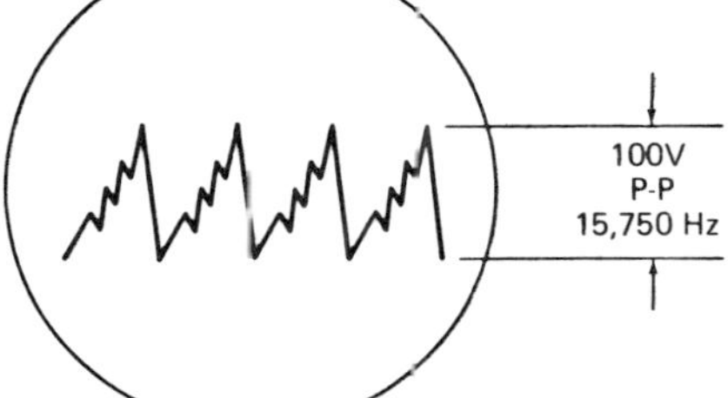

Fig. 6-1. This high-amplitude waveform was taken at the 21LU8 plate.

suspected. The scope waveform verified the conclusion that the circuit *was* oscillating. Bridging the capacitor cleared it up, so that part was okay. Now I knew what I was looking for.

An open capacitor? No. One with a very high power factor, for bridging a new unit across it did not cure the symptoms, but did *help*. This action is almost a sure clue to the presence of an electrolytic filter with a most extremely high power factor.

From the schematic, I saw that the output stage was fed from the 270V line. After tracing this out we found that it came from the 275V line, through the 10-ohm vertical centering control, and that there was only one filter directly connected to that point, a 250 μF at 350V marked C512A on the schematic. So this was my most likely suspect.

Now, back to work. I found all the filters mounted on the front apron. Hooking up the capacitor tester, I checked them all one at a time, and found a big triple unit that showed no response at all; this was evidently C512.

An ohmmeter check of the wiring showed this was correct. So I took all of the wires off, and rechecked; This was it; I could get no eye opening at all, even with the checker set up to 65% power factor, as high as it would go! So, there it was; diagnosis confirmed. I smiled happily, and went and got a cup of coffee.

Using Hindsight

Did you get it as it happened? Let us analyze the situation, using our well-known 20-20 hindsight. The negative voltage on a plate that should have had 250V or more positive told us that one stage was in violent oscillation; Fact 1: the hum and motorboating in the sound confirmed this. Fact 2: the high-amplitude waveform with a fundamental around 15.750 kHz gave further confirmation. Fact 3; bridging with a good unit helped but did *not* cure. This confirmed that I was looking for a bad capacitor and also told me what the defect was; high power factor, not a completely open one. Actual tests after pulling the chassis confirmed this.

I measured a "Do Not Measure" test point because I knew that there was *no* high-amplitude 60 Hz spike present! One look at the scope screen told me that. So I checked, and found the negative voltage. That, plus the hum (and later, the scope checks) told me that I did indeed have a case of violent oscillation. Experience told me that the most likely cause was a bad filter, and the character of the responses showed that it was apt to be one with a very high power factor, and it was.

Someone in the back row wants to know about the burnt-up resistor in the cathode circuit? That is apparently one that the set threw in to fool us. Just like a child's riddle. Evidently, when the oscillation began and the vertical output stage lost its normal drive signal, it drew so much plate current that the resistor overloaded and burned up. If the set was left in operation long enough, this heavy current could have damaged the tube; so I checked very carefully to see that it was okay; by checking the raster, linearity, amount of overscan obtainable, etc. If anything had shown up here, I would have replaced the 21LU8 tube before taking it back.

THE TROUBLE TREE

I've been over this before, and I'll probably go back over it again—and again and again. Not only for the benefit of the new men who come into our bewildering profession each year, but for the benefit of a few forgetful old goats like myself.

It deals with the most important thing in any electronics repair job—the diagnosis. Test equipment does not make diagnoses—your brain does. All the test equipment does is give you a set of facts. *You* make the diagnosis from interpreting these.

The first thing you need to keep in mind is that in a complex device like a TV set, or even a radio, there are as many potential causes for trouble as there are parts. In multiple-function devices such as color TV, this is even more important. So when we see a given symptom, we must make up a mental trouble tree, like Fig. 6-2. At the bottom is the major symptom, and the branches are all of the things which could possibly cause it.

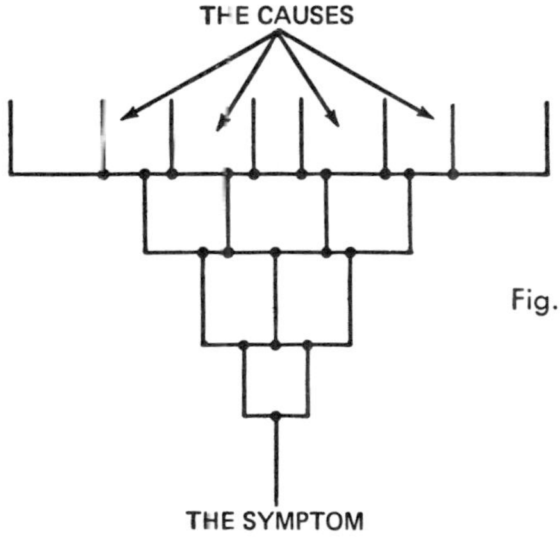

Fig. 6-2. The trouble tree.

A single, simple symptom can have multiple causes—for example, "no high voltage." A complaint of "no vertical sweep" is only a little less complex. Even "no sound" can have quite a few of possible causes.

Try growing one yourself. Sit down with a piece of paper, pick a symptom and then make up a trouble tree that names all of the possible causes. All, I said! Now, to make a useful diagnosis, start eliminating things one at a time, by testing.

To make things go faster, put the most likely or most common things first. This is easiest, and has the greatest statistical chance of success. For example, in a tube or hybrid set, experience has taught us that tubes are the most likely cause. They also plug in. So the logical choice for the first one would be replacement of the tube.

Next would be the dc voltage supply, and so on down the line. Now, having gone this far, get out the schematic of a set, and look at it. See if you can find something you've overlooked. Remember, every part is suspect! (How many of you put down "crack in PC board conductor" as a possibility? See what I mean?)

With a good knowledge of how the basic circuit action works, you can take shortcuts. For example, take "blank screen; no sound." Here, you have the whole signal path of the set as a possibility. Since this is a long circuit, break it in the middle. Take a scope reading at the output of the video detector. If you see a normal video signal at this point, you've eliminated all of the stages "to the left" of this point; tuner, i-f agc, etc. The trouble lies in the video amplifer, picture tube, etc.

If you find no signal at the video detector output, you know that the trouble is ahead of this point. Either the tuner, the i-f, or the control circuits. For another shortcut, clamp the agc with a bias box. It sets the control voltage at maximum gain (use the no-signal agc voltage shown on the schematic). If this produces a picture, you've cleared the signal-path stages and can proceed to the agc circuitry.

If the agc clamp does not produce a picture, leave it there. Start signal tracing through the i-f to find out how far the signal goes before it stops. By feeding in a high-level, fixed pattern signal, you can read its amplitude with a crystal-detector probe on the scope. The color-bar output from a bar/dot generator is very good for this, in color or black & white. In most sets, you can even see the small i-f signal output from the tuner, on the i-f input. This clears or confirms the tuner as a suspect.

A lot of technicians place too much reliance on dc voltage readings. They're essential, of course, but *should never be used as the only diagnostic tool*. From a lot of experience and troubles, I believe that the fastest way is a combination of tests: first, for the dc voltage readings at key points in the suspect stage. In other words, if you get a signal through the first i-f stage, you don't *have* to take dc voltage readings; they're good. Find the stage which doesn't pass the signal, and read dc voltages there.

No vertical sweep: Touch the scope probe to the grid of the vertical output tube. If you get the normal sawtooth signal there, the oscillator is running. Now, you've established that the oscillator is working, yet there is no vertical sweep. Having eliminated the most likely suspects, go to the yoke tap of the output transformer. If you see the normal spike signal there, the output transformer is cleared.

Now you have one suspect, or do you? Open yoke, yes. Also! Broken wire in yoke cable, bad contact in yoke socket, bad contact in yoke plug! Don't overlook the little things.

The key point I hope I made in all of this is that each test must be made for a purpose. Don't just sit there and poke at it. When you pick up the probe of a test instrument, know what you want to test and go to a test point that will tell you.

CASE HISTORY

This is the tale of a service job. Nothing really unusual about it, except for one mildly odd reaction. That's not why I want to talk about it. Rather, I want to use this to illustrate how the most "unusual" problems can be tracked down by the use of a logical series of tests, a good knowledge of the circuit actions and reactions (some of them exceedingly basic), and a plentiful supply of that quality known as "admirable determination" in our own kids and in those of the neighbors. Ordinarily, you won't have to go all the way through this whole series of tests. In this case, I did. That's the interesting part.

The subject was an RCA CTC 31 color TV. The raster came on, perfectly focused, good color, and plenty of brightness. It stayed on for *exactly* one minute from the time the picture appeared. Then, it dimmed, lost focus, pulled in from the sides, and disappeared. Turned off for not more than 30 seconds, it would faithfully *repeat* this, indefinitely.

Now here we go. Preliminary readings showed that both high voltage and focus were dropping about 40% when the raster went out. Monitoring the 6JE6 cathode current showed it stayed normal at 200 mA with raster, but it dropped only about 10 mA when the raster went out. The scope showed full drive on the 6JE6 grid, and the grid voltage was fine. Neither of these changed at all when the trouble occurred.

Taking the easiest and most likely things first, I changed the horizontal output, damper, high-voltage rectifier, focus rectifier, and high-voltage regulator tubes. No help. Then I substituted deflection yoke and picture tube, with a test jig. Same thing. From here on, I went through a series of tests. In each one, the result was exactly the same; no change in the symptoms or the timing of the raster fading. Very uncanny! Here are the tests:

1. **Replace** tubes: horizontal output, damper, high-voltage rectifier, high-voltage regulator, focus rectifier.
2. **Monitor** 6JE6 cathode current.
3. **Substitute** deflection yoke and picture tube using a test jig.
4. **Check** deflection yoke socket, wiring, plug.
5. **Scope** 6JE6 grid-drive waveform; read dc grid voltage.
6. **Disconnect** focus transformer from flyback.
7. **Substitute** horizontal efficiency coil, capacitors.
8. **Read** 6JE6 screen-grid voltage.
9. **Read** high-voltage regulator cathode current.
10. **Substitute** high- B+ rectifier.
11. **Substitute** 0.0022 μF capacitor in high B+.
12. **Check** screen controls, picture tube.
13. **Read** screen current.
14. **Check** vertical linearity control on high B+.
15. **Disconnect** vdr in pincushion circuit.
16. **Check** arc-gap capacitors, screen circuits.
17. **Spray**-cool *everything* in sight.
18. **Read** picture tube cathode voltages.
19. **Read** picture tube grid voltages.
20. **Read** color-difference amplifier tube plate voltages.
21. **Read** dc voltage on kine bias control.
22. **Check** difference amplifier tube heaters.
23. **Read** blanker tube plate voltage.
24. **Read** blanker tube cathode voltage.
25. **Check** blanker tube.

Substituting a new 6GH8 for the blanker tube stopped the trouble. Putting the old one back made it go back to its one-minute dropout. I put the tube in the tube tester. It would go up to full emission; in one minute it dropped to zero. I repeated this test four or five times. This thing would make a marvelous electronic eggtimer. In the meantime, the set was playing merrily away without a bobble.

A very close examination of the tube with a high-powered glass showed an almost imperceptible lump on the cathode ribbon, from the base pin to the lower end of the cathode cylinder. Another tube didn't show this lump. It didn't look like a bad weld; I decided that it was a fatigue-crack in the ribbon, which opened up when the cathode had been heated for precisely that amount of time! Cooling, it closed again.

So, there it was. Elapsed total time, something like 2.5–3 hours (with, I will admit, numerous coffee breaks!) Explanation: when the blanker cathode opened, it opened the dc path to ground for the circuit which determined the picture-tube grid voltages. When they went highly positive, the picture tube drew far more than normal beam current. This overloaded the HV supply and pulled the voltages down to the point where the screen went dark.

This was a good fight, and I enjoyed it. It was also a lot of good practice in diagnosis! The checklist might be useful. It could help novice technicians remember how many things you have to check to give a TV set "the full treatment!" Just remember what each action and reaction *means*, and what it's trying to tell you. With a large supply of patience, that's all you need!

RECOGNIZING INTERMITTENTS

Before closing the door on our discussion of *diagnosis* as an aid in troubleshooting, let's have a few words about intermittents. (Not the words normally used around the shop; this book must go through the mails.) But the truth is, intermittents make up one of the largest stumbling blocks to effective diagnosis and efficient troubleshooting.

An intermittent is just that; something which makes intermittent *contact* in a circuit. It can be a short, but most of the tougher ones seem to be opens. They can be broken down into two general types—the *physical*, or "jar" intermittent, where you can make it act up by tapping the chassis; and the *thermal*, caused by heat.

The jar intermittents are much easier to find. By tapping the chassis very carefully you can find the most "sensitive" area, and from this point it's not too hard to locate a guilty part or solder joint. The eraser of a lead pencil makes a dandy tapper.

Thermals are tougher. Even in a transistor chassis, you're going to find thermal intermittents. This is especially true of those long-term intermittents that cut out once a day. Something in there is getting hot and expanding. The circuit opens (or shorts to other parts) and there you are. In the standard thermal, no amount of jarring the chassis (or kicking the cabinet) will faze it.

There are a couple of tricks you can play on these. One is running the set with the ac line voltage just a little above normal. This won't hurt anything unless you leave it on for too long. The high line voltage will raise the dc voltages. It will also raise the temperature of things and make the intermittent more apt to act up. If this doesn't work, try running it with line voltage a little below normal. This leaves things a little cooler than they usually are.

Determine the characteristics of the trouble. Does it cut out once a day, every 2–3 days, every hour or so, etc. Questioning the owner is a very good way to get this information. If this is a long-term intermittent, and irregular, the chances are that it is thermal.

There's a division here. If the intermittent shows up within a short time after the set's turned on, say 5 to 10 minutes, the chances are that it is in some part which generates its own heat; a current-carrying resistor and so on. Solder joints are frequent offenders in this category.

The other and much more difficult type is the long-term intermittent—this shows up at irregular intervals, from hours to days. These are also thermal, but are caused by "conducted" heat, from other parts on the chassis getting hot. This heat is conducted through the chassis or the air. In "touchy" intermittents, the frequency of cut-out often depends greatly on the ambient temperature. (We'll show you a real wowser like this in a moment!)

In both types, heat is the troublemaker. It causes physical movement of the part, making or breaking contact. This can be as little as a thousandth of an inch, if it's inside a resistor, coil or capacitor; but it's enough! Printed-circuit board solder joints are frequent offenders. A "pocket joint" (see Fig. 6-3) can make perfect contact. But if the board expands slightly, whammo. If you suspect this, you can "get out the shotgun"; remelt and resolder *all* solder joints in the area of the trouble. You may never know which one it was but you'll fix it.

Isolation

The first thing you must do is isolate the area of trouble. From the characteristics, you know *about* where it is; that is, signal circuits, sync, sweep. Observing the characteristics will give you clues. Now, check every part of the circuit that could affect that particular function.

Now, see if you can make it show up. To do this, you can either apply more heat, or less heat—cool it. The reaction to

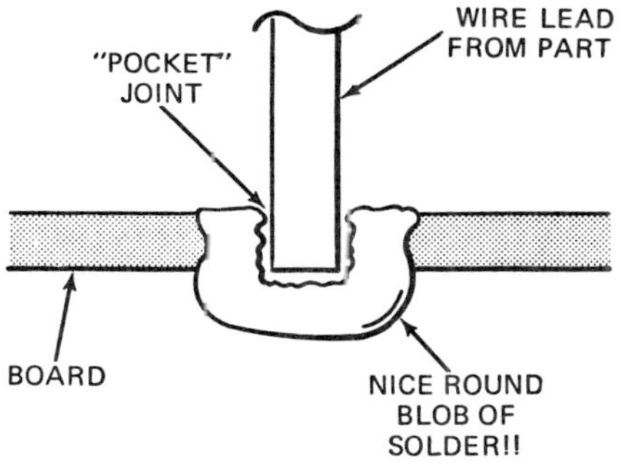

Fig. 6-3. "Pocket" joints like this are more common than you might suspect. The lead makes "touch" contact to the board until internally generated heat makes the board expand—and then the connection becomes an open!

these tests will give you a lot of data and that's what you need. You can apply heat with a heat lamp or by touching the suspect part with the tip of a soldering iron, or by blowing hot air on it with a heat gun or old hair dryer. Cooling is much easier. The spray-can coolant is a very useful weapon for locating intermittents.

Here's a good example. I had a Zenith color TV set on the bench. The picture flickered on and off at highly irregular intervals, leaving a white screen. This was a very fast flicker, sometimes repeated three or four times. After doing this, it would play for a couple of hours. In fact, it sat on the bench under observation for two days; no flicker! Fortunately for my production ratio, I was working on other things at the same time!

Then I got after it in earnest. Turning it on, I waited. This time, it started to flicker in only about 20 minutes. A quick check with the scope and a crystal-detector probe showed a steady signal at the i-f input, and a severe flicker at the video detector output. So, I had it pinned down; it was in the i-f.

Now, I made thermal tests. With a junk hair dryer which still worked, I blew hot air on the i-f stages. On the first and second stages, no result. As soon as I hit the last i-f stage and detector, wham! It started flickering like mad. To verify this, I blew spray coolant on it. The picture came back as steady as a rock. Now I had it!

The i-f stage in these sets can be taken out by pulling off some push-on connectors and taking out three screws. Making up a set of connecting leads, I put it on the bench and turned it on. Despite a tendency to oscillate, it worked well enough for testing. I checked each part in the output stage by touching it with the tip of a soldering iron. When I hit the output transistor, out it went!

Verifying this, I sprayed the thing with coolant. It came back on. Replacing this transistor cleared up the trouble. Hooking the original transistor to a curve tracer, it checked perfectly. Heating it with the soldering iron, in about 20 seconds it went to the horizontal line; it was opening up internally! Spraying it with the coolant brought it back again. I could repeat this as often as I wanted to. For a definite check, I warmed up my replacement transistor with the soldering iron. Absolutely no change in the picture.

About this time, someone will say: "But you'll ruin those delicate transistors, heating them up like that with a soldering iron." To which I can reply, "Possibly so; however, I have never damaged one yet with this method, and I've *tried*!"

I will admit that I've been guilty of saying that there was no such thing as an intermittent transistor. I stopped this; people kept sending me intermittent transistors, and I kept finding them in sets I was working on.

They *can* be thermal. In fact, a post-mortem analysis of the case of the Zenith brought out an interesting fact. The reason it sat on the bench and played for two days was that during those days the shop *air conditioner* was running, and blowing right at it! On the day that it started cutting out, the air conditioner was turned off! This comparatively small difference in the ambient temperature was enough to make it start acting up.

There are many ways you can go after this kind of trouble; don't neglect any of them. You can jar it, you can change the supply voltage, or you can heat it or cool it. Somewhere in this list, you should be able to find something that will make it show up. When you do, you've practically got it licked.

Varying the Voltage

Most of us say Variac when we mean any variable-voltage transformer, but **General Radio** might not like it, so I'd better use variable-voltage transformer or VVT, since there are quite a few different makes available now.

At any rate, thisthing can be one of the most versatile and useful pieces of test equipment in any electronics service shop. In fact, it could be called invaluable for work with solid-state equipment. Let's look at a few of the things it can do.

Here's one that comes in regularly in the mailbag: "I've just replaced all four big power transistors in this amplifier, and I'm afraid to turn it on!" That's a condensed version, of course, but you know—you've been there. So have I. Here's an easy way to avoid a heavy financial loss and some gray hairs, in cases like this.

Open the collector power supply circuit, or the main dc supply circuit, and hook in a suitable dc ammeter. Now, plug the thing into the VVT. Set this to zero volts and turn the equipment switch on. Now, start bringing the ac voltage up a little at a time, watching that ammeter like a hawk. Let's say that the normal resting current (no signal) on this amplifier is 1.0A.

When you get the line voltage up to around 50 volts, you see that you're already drawing 1.3A!

This is the time to back off! Stop right there and investigate. You have *still* got a problem! With symptoms like this, the most likely thing would be a bad bias diode, emitter resistor, or shorted driver transistor (in the common direct-coupled circuits). Something is playing heck with the *bias* on those four big (new!) output transistors. Since they're taking more rest current than they should, with the supply voltage only half normal, the chances are that they *would* blow out if you slapped the full line voltage on them without checking!

By reducing the input voltage, you reduce the dc supply voltage in direct proportion, in sets with solid-state rectifiers. Even with tube rectifiers, this method will still work. You can hold the *power* dissipation in output stages, or any high-power stage, down to a safe level without resorting to the famous *smoke* test.

In solid-state circuits, you can read the bias voltages, collector voltages, etc. and get an idea of what they would do at normal voltage. In other words, if the supply is half normal, the bias ought to be about the same percentage, and so on. The most important thing is the *current* drain, since this is the main figure in the I^2R power formula.

You'll be surprised; a great bulk of this type of equipment will *work* very well at greatly reduced supply voltages—well enough, that is, to make many useful tests; even distortion and that sort of thing. If it's clipping at these reduced voltages, it'll be worse at full supply voltage.

What if the equipment is dc powered? 12V automotive, 28V aircraft radio, etc.? Same thing. The only difference is that now you connect the bench dc power supply to the unit, and then plug the power supply into the VVT. This will make an ordinary dc power supply into a very handy, continuously variable dc voltage supply. Even a stock battery charger, with a filter capacitor across the output, becomes a useful dc source for anything from little 9V transistor radios on up.

You can start at zero and bring the dc voltage output up to maximum very smoothly. For cases where you want to check at higher than standard voltages, crank the VVT up to 140V line, and your 12V supply will go to 14 volts and more. This is the exact voltage you want to check things, such as simulating actual operating conditions in a car with the engine running at cruising speed.

TV sets? Same thing. You've just replaced the horizontal output transistor, or some other high-power device. Can you turn it on without wincing? Yep. Plug it into the VVT, start at zero volts, with the current meter in the circuit, and sneak up on it. Here again, you'll be surprised at how well the thing works with far less than the normal ac line voltage. I've seen quite a few solid-state TV sets that would make good pictures, with ample width and brightness, at line voltages as low as 90V.

Still in the TV department, this thing is ideal for those weird complaints of loss of width, brightness, etc. at the home, although the blasted thing will sit on your bench and play perfectly. Plug it into the VVT, and run the line voltage down a little. See how far down you have to go before you see a loss of width, or whatever it was. This will give you some idea of the condition of things. For a rough rule of thumb, the average set should go down to at least 100V input before you lose too much width, etc. (Ratings on the label are 105–120 volts.) If it'll do this, the set is probably okay and you could have an intermittent drop in the line voltage at the home.

Another handy use: checking for intermittents in the same class; "Won't play for more than half an hour at home. Plays all day on the bench!" Plug it into the VVT, crank the line voltage up to about 125 or 135V, and let it play for about an hour. This will help show up obscure troubles, such as transistor leakage due to excessive collector voltage, heating, and so on. To the best of my knowledge, this will *not* damage a properly designed set in good working order. At least, I've used this test on a lot of sets, and never blown up a good one yet. The point is this: if the intermittent is thermal in character, a little extra voltage will move the problem from the *intermittent* category to a stable—and therefore easily repairable—condition.

You'll find dc voltage regulator circuits in a lot of things lately. The VVT is ideal (essential is a better word) for checking these. You will even find voltage regulators in a lot of units used on battery power, such as car tape players, etc. By using the tests given just now, both for ac- and dc-powered things, you can run the input voltage up and down at will, while monitoring the voltage-regulator output.

Heat-Up Distortion

I've got a stereo amplifier that distorts badly after it's been on for about 5 minutes. I replaced some of the output transistors, and I've checked the heatsinks to make sure that none of the leads were shorted. What could cause this?—L. J., Little Neck, NY.

Two things. This could be an overheating condition that slowly warms up the junctions of the power transistors and starts them clipping, etc. This could be due to insufficient heatsink area, allowing the cases to get too hot, and the junctions with them.

There are two possible cures for this. One is larger heatsinks, or the addition of a small forced-air fan to the chassis. Might try spraying coolant on the transistors after the distortion shows up, and see if this clears it up. If so, better cooling *would* be the answer.

For the other, I would recommend replacing all of the original transistors.

In some cases, you might get rid of this trouble by using a slightly *larger* transistor. Many of the import amplifiers use the small "cigarette-filter-sized" output transistors, and run them hot enough to smoke! Replacement transistors such as those mentioned above, and others, have larger cases, and can be run at a higher level of power dissipation without overheating.

Check all transistors for leakage. This can cause severe distortion, and overheating makes it get worse.

Intermittent Flare, Raster Collapse

In a Clairtone C11 color TV chassis, the picture bloomed, brightened, showed retrace lines, then faded and went out, at intervals. I replaced the 3A3, and it cleared up for about 10 minutes, then went out again. Pulled chassis, took to shop. As expected, it played for hours!—S. J., Scarborough, Ont.

I started to answer this, then got another letter! In this, the reader said, "Never overlook the too-simple and obvious things! While waiting for the thing to cut out, I pushed down on one corner of the PC board, and the raster flared and went out. So did the *heater* of the G−Y amplifier, the R−Y/B−Y amplifier and the X demodulator. Common ground point for these heaters is an eyelet on the corner of the PC board. Now everything is fine.

S-Bending

I have a very bad case of bends in the picture, with sync instability, etc., in a GE AA chassis. I have checked all of the filter capacitors, with no results. What is causing this?—C. L. E., Malacanang, Philippines.

This chassis has two main PC boards, mounted to a sheet-metal frame with self-tapping screws. If either of these boards develops a high-resistance ground connection, you'll get symptoms exactly like this.

Suggestion: loosen and retighten all of the grounding screws. If this doesn't help, then run extra grounding wires or shield braid between the ground foils and the chassis. Most likely suspect: the one with the video i-f, agc, and so on.

Intermittent Roll

I'm working over an old Zenith 14N28 chassis. Got everything going fine except for an intermittent vertical roll. Checked all capacitors and resistors in the vertical oscillator, changed tubes. Any ideas?—W. M., Lagrangeville, NY.

Check the two integrators in this circuit; these are the little "three-legged" types, looking like dual ceramic capacitors. I doubt if they'll be shorted, but you may find the one in the feedback loop has increased in resistance, or is intermittently opening up. Normal resistance should be about 87,000 ohms for the one in the sync, and 90,000 ohms for the feedback-loop integrator. If you can't get the Zenith parts, a Centralab PC-407 or PC-408 will replace them.

Floating Blur

I've got an odd floating blur in a Philco 19KT51 chassis. It's about 6 inches high and floats slowly up the picture. When the blur is at the bottom of the screen, I can see video that should be at the top of the picture, and vice versa.—A. M. Philadelphia, PA.

Whatever it is, it is a 60 Hz pulse (only one bar visible). Suggestion: Put a scope on the dc supply lines around the video stage, and look for anything like a 60 Hz ripple or hash, pulse, etc. A heater–cathode short in the video output tube will cause the same kind of symptoms. I've seen an rf amplifier tube cause similar symptoms.

White Flashes

I've got an old RCA CTC 9 color chassis on the bench. Works like a charm, with a few new tubes and a resistor or two, and realignment. But the set has an intermittent white flashing in the picture. The whole picture will go very bright, and out of focus a little, then go back to a good color picture. Long-term intermittent. I'm afraid to take it back until I find this—L. H., Honobra, OK.

This is caused by something that is affecting the video (Y) signal. The pix tube cathode voltages are going negative (which makes the picture tube go to higher brightness) then falling back to normal.

The most likely cause for this is an intermittently leaky or shorted tube. Try a new one here. While you're there, clean the socket contacts and check all of the PC board conductors for breaks.

Vertical Blink

I've got an Airline color TV in for service. The raster will collapse to a very thin line, then slowly fill up again. Action is pretty fast. This is a GEM-17148B chassis, and I can't find a listing in my file. The vertical tube (13GF7) is okay, and the dc voltage supply is okay.—R. D., Vinita, OK.

For a schematic, check the **Sharp CN-52T** which is practically the same, especially in the vertical circuit.

Check the cathode voltage on the vertical output; pin 3 of the 13GF7. If it is jumping up and down, turn it off and check for continuity to ground. Should read 1470 ohms; a 470-ohm resistor on the PC board near the socket, and a 1000-ohm resistor (a 2-watt film type). The last one is probably opening intermittently.

By the way; this 1000-ohm resistor is completely inaccessible unless you pull the chassis. It's in a group of similar resistors to the left and below the opening you see by taking off the bottom plate. Trace the circuit from the 13GF7 socket to the 23JS6 (horizontal output tube) socket. You'll find a thin green wire going to pin 4 (suppressor) and then going on over to the 1000-ohm resistor. Quick repair: Just clip the wire to the original resistor, and solder a 5-watt wirewound 1000-ohm resistor from the lance on the PC board (bottom of the 470-ohm resistor) to ground. There is plenty of room for this in here. Use spaghetti on the leads.

Intermittent Brightness

I thought this one would be simple but it wasn't. In an RCA CTC 15, the raster would go out. I moved the 12BY7 video output tube and it came back. I replaced the tube; got about a week, then the same thing. So I replaced the tube socket, and I've still *got the same problem! The 12BY7 goes dim and out goes the raster.—N. P. S., Greencastle, PA.*

I've had the same thing in my own ancient RCA. Run a *jumper wire*, solid No. 20 at least, from the ground terminal of the 12BY7 heater to the nearest ground. Now, for luck, run another from the hot 6.3V socket terminal to the nearest lance on the board (toward the front of the video board, brown wire).

I think this will cure it. There is a hairline crack *somewhere* in the heater circuit, and this is the fastest way to get it.

Finding Shorts In PC Set

I've got a Magnavox 1-T-109, with PC *boards; it had a short in the B+. I've changed the rectifier and filter capacitor, both shorted. Now, I read 20,000 ohms to ground on the B+. I think this is too low. How do I get into this type of circuitry to separate the loads?—T. P., Brooklyn, NY.*

Actually, 20,000 ohms to ground isn't too bad on a +145V supply. Try reversing your ohmmeter prods and see if it isn't higher the other way. If so, okay. Install the new filter capacitor and rectifier, but leave the cathode lead of the diode open. Connect a 0–500 dc milliammeter in this lead, between diode and filter.

Now turn the set on and check for excessive current. Normal should be about 300 mA. There's a sneaky you can pull; just tack a 300 mA pilot light in series instead of the meter. If the bulb lights to full brightness, okay, if there's a short, it'll blow out.

Intermittent Sync Problem

When my customer changes channels on the Sylvania D14 chassis, the set flips its wig! Horizontal sync drops out completely. I tried to compensate by adjusting the agc, which helped—but then the problem changed to vertical jitter. I've heard about a factory modification to prevent this—do you know what it is?—K. S., Burlington, KS.

Yep. Sylvania says: Intermittent loss of horizontal sync when changing channels may be a complaint on early chassis

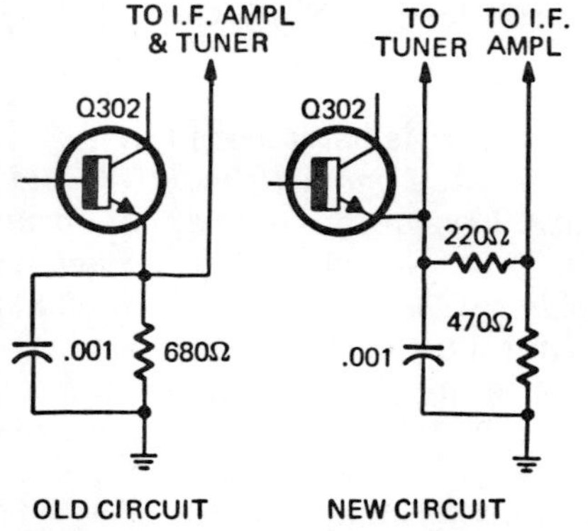

Fig. 6-4. Factory modifications to Sylvania D14 chassis to improve stability.

without a video buffer stage. All checks on the horizontal and sync circuits show normal operation. Advancing the agc to the point just before overload corrects the horizontal problem but results in vertical jitter. Modifications can be made to eliminate these problems:

1. Change the emitter resistor in the first i-f stage (Q200) from 15,000 to 8200 ohms.
2. Modify the agc amplifier circuitry as shown in Fig. 6-4.
3. Change R334 in the base circuit of Q308 from 470K to 390K. Any lower value than this will affect horizontal phasing and can cause excessive horizontal jitter on weak stations.
4. Change agc filter capacitor C304 to approximately twice its original value. Too much capacitance will cause airplane flutter and increase the affects of cochannel interference.

Helpful Hint, Admiral Color

Some time ago, you replied to a reader about an intermittent short in Admiral color chassis such as the G13, H10, or H12. Here's another good possibility:

In the non-instant-play versions they use a thermal switch. Part 77C160-3, for the degausser timing. Some of the turns of the heater winding will short, causing the bimetal blade to overheat and warp. When the set cools, the contacts will open normally, but the warped strip keeps on its merry way and grounds *to the case!*

So when you turn it on again, the dc power supply is shorted to ground while the strip is heating! After a few seconds have passed, the strip heats, the short clears, and everything is fine. It will operate normally once the circuit breaker is reset.

The cure is to replace the thermal switch. Use the later version, 77C160-4. Follow the instructions with it and tape up the heater lead.—Charles E. Koontz, Burlington, IO.

Chapter 7 TV Sets and CATV

Not all the problems we see can be traced to the TV's innards. And not all of the externally caused problems are the result of power-line arcing or other interference sources. Two good examples of picture-problem sources—particularly troublesome in low-budget cable installations—are videotape recorder anomalies and CATV imperfections.

CABLE SYSTEMS

Like everything else, cable systems have their peculiarities. The first question, in most cases, is: "Is it the set or the cable?"

With the right tests, it's easy to find out. If the set seems to be in good working order, but there is no picture, a weak picture, etc., this is the point where the correct test methods can save a lot of work. It's no fun to haul a set all the way to the shop only to find it plays up a storm on your antenna.

The quickest and most definite test is a substitute signal. A bar/dot generator is ideal. It will tell you whether the set is working, whether it will make color and sound, etc. For the best results, measure the rf output of your bar/dot generator on a field-strength meter, if yours has a variable rf output control. If it doesn't have a variable rf output, find out what the output actually is. In a cable-system, *each* channel should have a signal strength of about 1000 mV. This will make a snow-free picture on any TV set even in fair working order.

You'll find some problems that are peculiar to CATV systems, if there actually is a fault in the system. One typical trouble is the case where the high channels will come in clearly, but the low channels are very weak. Or, vice versa. Or, all channels are weak and snowy. First check, hook up the bar/dot and see if the set will make a clean picture. If it will, then you have "cable" trouble.

The most common cause of this is a defective balun at the cable outlet. If it has been visited by lightning, it could be partially open somewhere. Depending on exactly where it is open, you'll get this kind of symptom. By the way, you can usually check high and low channels on a set, even with a fixed-tuned bar/dot generator. For example, if yours is set on channel 3 (60–66 MHz), you should be able to see a pattern on channel 8 (180–186 MHz) on the 3rd harmonic. It will be only $^{1}/_{3}$ the amplitude of the channel 3 fundamental, of course, but enough to make a picture.

A field-strength meter, of course, will give you a definite answer as to signal strength on each channel used. If you get a reading considerably less than 1000 mV on any or all channels, *then* you call the cable company's technician, and tell him what you found, and why.

Ghosts on Cable?

It *is* possible to get ghosts on a cable system. If any of the coaxial cable feedlines aren't properly terminated, you get standing waves. One cause of this is accidental damage to the coax during work on something else in the building. However, a properly designed and installed cable system will be free of ghosts. For a check, get out the bar/dot generator again. See if you can tune in a crosshatch pattern; this will tell you whether the ghost is in the set itself. Tune for the sharpest vertical lines in the pattern. If there is any tendency to ringing in the set, this will show it up.

Color Loss

I have heard complaints that "the cable kills the color." On the face of it, this is mildly ridiculous. The line amplifiers are extremely broadband. To let the video signal get through, and trap out only the color, the response-curve of the line-amplifier would have to have a very sharp "notch" in it, exactly on the color subcarrier frequency. This is impossible. (I know; we tried to *make* one do it, some time ago, and couldn't.) It may be possible; but all I can say is that I have never yet seen a verified case of it.

It *is* possible for line amplifiers to add hum bars to the rf signal. In vacuum-tube line amplifiers, heater-to-cathode leakage in a tube will do this. Sometimes you'll see quite a bit of bending in the picture, too. An open filter capacitor in the dc power supply of either tube or solid-state line amplifiers will also make hum bars. To find out whether it's the set or the cable, haul out the bar/dot generator again. Put a crosshatch pattern on the set. If the hum is in the set, it'll show up as background shading, or bending of the vertical pattern lines.

Some TV sets will have agc problems on cable systems. These are mostly the cheaper models, without an agc adjustment. Symptoms will be standard; overload in the picture, bending, buzz, and so on. For a quick check on this, disconnect the lead-in and hold it close to the antenna terminals. If you can find a point where the set will make a good clear picture, but it buzzes and overloads when the lead-in is hooked up, that's it.

In some cases, you can add a simple resistive attenuator to bring the signal down to a range where the reduced agc can handle it. Sensitivity of the set can be roughly checked by disconnecting the antenna. If you see a good deal of snow, the rf gain is usually okay. If you see a nice clean snow-free raster, look out. This normally means a loss of gain; something like a weak i-f tube, etc.

Crosstalk and Cable Bars

In an improperly adjusted CATV system, if the signal levels are too high, you *can* get crosstalk between channels. This will show up in most cases as floating blanking bars on the screen; beats, ghosts, etc. If you see this, check the signal strength meter. You'll probably find some channels running much too high. If your bar/dot generator will make a clean picture, then it's time to ask the cable technician for help. All properly designed CATV systems have an agc control at the head-end. amplifiers, and many use agc on all amplifiers, just for this purpose.

Cable Bars on Antennas

There is one more. Not too common any more, though we used to run into it now and then. This is the appearance of a white or black blanking bar on the screen of a TV set working on an outside antenna, but close to one of the line amplifiers of a CATV system. The cause is simple; the line amplifier is leaking rf signal. When the set, on the antenna, is tuned to a channel being carried by the cable, it picks up the direct signal from the TV station. It also picks up the rf leakage of the same signal from the line amplifier. Since this signal will be *delayed* from coming all the way down the hill from the head end, the blanking bar will show up in the picture. (Velocity of propagation in coax cable is about 35% slower than in free space; delay). To find out, aim the antenna directly at the suspected line amplifier, which will be on a nearby pole. If the interference gets worse, that's it.

This can be cross checked by asking any of the neighbors who are also using outside antennas. If they all get the same kind of interference, it's time to call the cable technician and give him the word. This is usually due to someone leaving the lid off or loose on the line amplifier housing. (The FCC has set up a standard maximum for rf leakage from cables and line amplifiers to prevent this kind of interference.)

THE VIDEOTAPE MACHINE

To a reader's question, I answered (slightly edited): "Those odd horizontal streaks you see in the picture are due to a problem in the videotape recorder (VTR) at the TV station." Whereupon, the roof fell in on me! I got letters, long distance calls, etc. from irate VTR engineers, and even from the Telco people. All of them said, in effect, "You're crazy! A *properly operated* VTR will never do that!"

To be as plain as possible, that is exactly what I was saying; a properly operated VTR won't do it. A dirty, improperly adjusted, incorrectly operated VTR definitely *will*! The next sentence in most of these letters was "You'll never see this kind of trouble on a TV station or network. The FCC won't allow it!" Right again. However, I *have* seen every one of these troubles, not only on local TV stations but on certain network shows, with my own bloodshot baby-blues! Loss of color sync and horizontal sync, a complete wipeout of the picture, as well as the "streak every 4 inches" symptom!

I did not correctly identify the precise VTR troubles that caused these symptoms. Much detailed correspondence with some very helpful and highly competent people helped to clear this up. Chief among these were Mr. James O'Neal of New Haven, CT. a VTR engineer, and Mr. Eric Address, chief engineer of E.J. Stewart, Inc., of Broomall, PA. Mr. O'Neal simulated all of the errors, photographed them, and allowed us to use the pictures.

Distortions and Their Causes

The sawtooth look in a picture should be called "skewing." Figure 7-1 shows the unmistakable symptoms. The distortion results from improper tape-to-head contact. It used to be quite common in the early machines. Now, it's automatically corrected, by special circuitry, which makes a correction for *each* horizontal line! Figure 7-2 shows the error signal developed by the time compensator for the pattern of Fig. 7-1; this was disabled in order to misadjust the machine to make the photo. The late models also have an automatic guide servo to set the "head penetration" or guide position automatically.

Figure 7-3 shows a color-bar pattern with a very severe velocity error, deliberately introduced. Here, the color bars

Fig. 7-1. Videotape playback with a large amount of "skewing" geometric distortion. The video signal feeding the monitor in this photo was obtained at a point in the tape system before any time element compensation had taken place. (Courtesy J. O'Neal.)

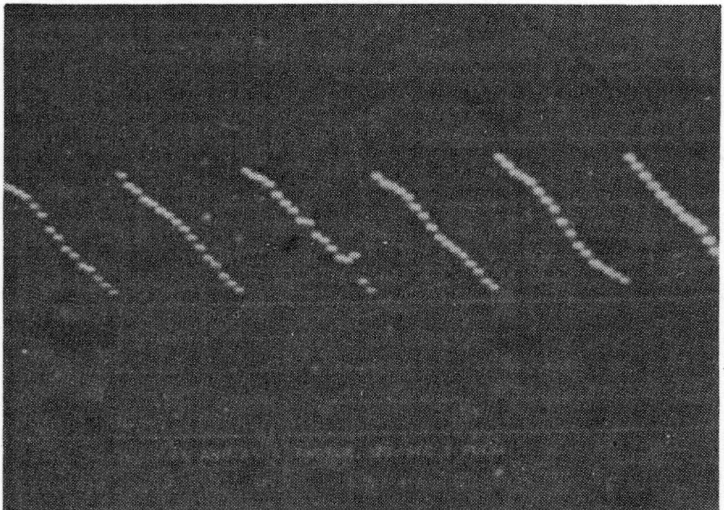

Fig. 7-2. Oscillogram of error signal developed by the time element compensation device for correction of skew shown in Fig. 7-1. (Courtesy J. O'Neal.)

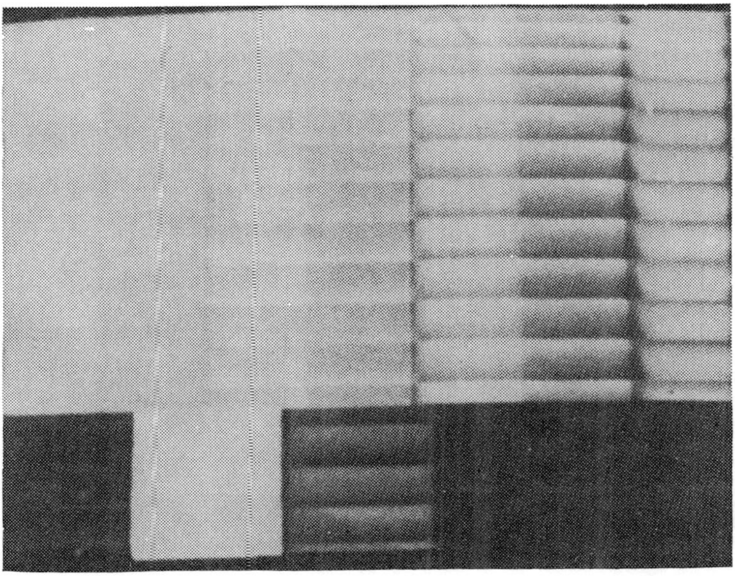

Fig. 7-3. Extremely severe "velocity error" introduced into the tape system for photographic purposes. Individual color bars are straight even though a severe color shift is occurring. Note also the absence of error in the first "gray" bar and white 100% area at the bottom. (Courtesy J. O'Neal.)

are straight although a very severe color shift is taking place. Note the absence of error in the first gray bar (left) and in the reference-white area below.

In Fig. 7-4, we finally see a very good example of unequalized-head error. Actually, Mr. O'Neal had a hard time getting this photo! The head itself isn't bad; he finally had to remove all of the high-frequency equalization from his preamp. The resulting HF loss dropped the chroma level and gave a perfect simulation of this type of problem.

I will once again agree with the gentlemen who properly called me to account on the first item. You won't see these types of problems in a properly operated clean VTR. You may see them now and then; more frequently if you watch some of the smaller TV stations.

These conditions are almost always intermittent. They can be the result of minor errors; VTR not quite up to speed on cue; momentary loss of head contact, etc. In the larger TV stations and networks, a backup (duplicate) tape is kept synced and rolling on another VTR! If any trouble occurs, this

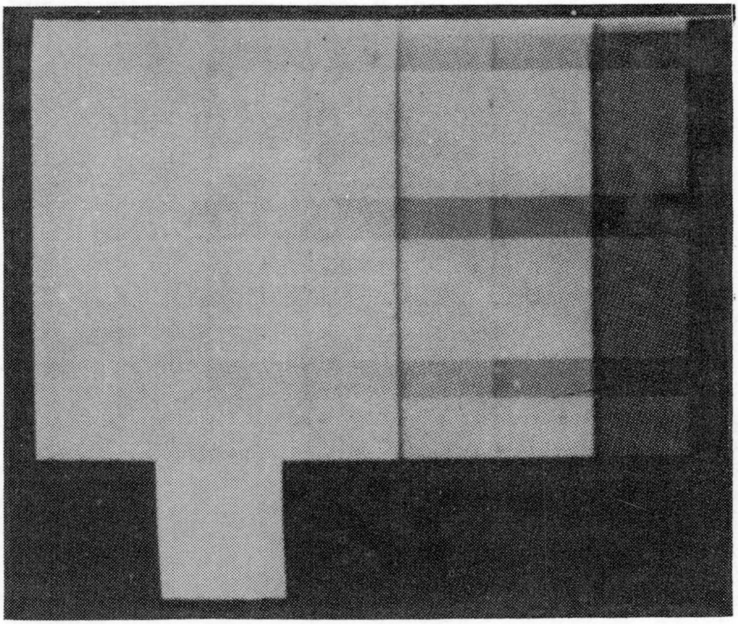

Fig. 7-4. Banding distortion produced by a bad head or unequalized head. (Head was actually in good condition. All high-frequency equalization was removed from head preamp and the resulting high frequency loss dropped chroma level.) The wide diagonal band was produced by the camera shutter and not visible on the monitor. (Courtesy J. O'Neal.)

is instantly switched in, so fast that viewers often cannot see it!

There may be "emergencies," too. I once watched a whole hour of tape, with a complete loss of color sync and lousy sound! However, this was the only recording available of the college football game, so they went ahead with it (and then ran for cover!)

Black "Side Streaking"

While we were at it, Mr. O'Neal cleared up another very common reader question—the "black streaks to the right of light-colored letters on a dark background." If this is not present at all times in the TV receiver, or on all stations, this can be due to video problems at the TV station. The major cause is very poor low-frequency response *somewhere* in the TV system. The most likely cause is misadjusted frequency compensation in the camera-tube pickup preamplifier. Partial or complete loss of clamping in any clamped amplifier stage can do it.

The "white-to-black" (or vice versa) transition in a picture should be a prefect square wave, with very short rise and fall times. Figure 7-4 shows a square wave with negative undershoot, also very poor low-frequency response, as shown by the overshoot and tilt on top. This type of response would produce the trailing smear symptom in a picture. In highlight areas, this isn't noticeable; in any scene with an abrupt transition from light to dark, or vice versa, it is.

However, neither of us is absolving the TV receiver from possible guilt. If some of the low-frequency compensation in its video amplifier is defective, it can definitely show such symptoms. If this is true, then it would show up on a fixed pattern, such as color bars, "shading patterns," and so on (unless, of course, it was intermittent!). In any case, the receiver should be thoroughly checked out before blaming the TV station for all troubles!

I reiterate that I am not making a blanket condemnation of TV operating engineers! The worst thing I said was that some of them may not have enough yet. It isn't hard to make mistakes. I know of one case where a radio engineer read almost half of a 15-minute newscast into a very dead mike, simply because he forgot to push the mike switch down! I wouldn't mention any names, of course. (At least, not any of the ones the station-manager called me when he came running in screaming, "You're off the air! You're off the air!")

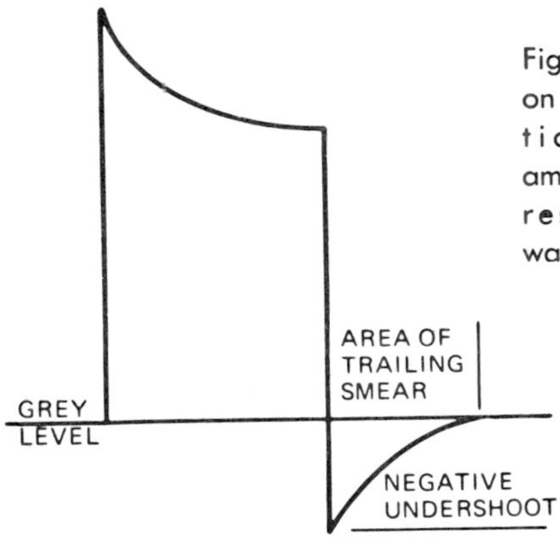

Fig. 7-5. Trailing smear on black-to-white transition is caused by amplifier with this type of response to square waves.

MATV Problems

I'm overhauling an old installation of an MATV system and I've got troubles. There are 4 antennas feeding a mixer amplifier, which in turn feeds a broadband line amplifier. This feeds a distribution system with 13 outlets. Stations are 30 miles away. The pictures aren't good enough; quite a bit of snow, and weak sync, etc. How can I get a better picture?—S. K., Washington, PA.

Signal-trace it, and I mean that quite literally. Get a field-strength meter and start at the antennas. At this distance you should have something like 400 to 500 μV minimum on each antenna. Now check the signal level at the mixer amplifier output. Should be quite a bit more, around 1000 μV per channel.

Then check the signal at the output of the line amplifier. You shouldn't have too much loss unless your interconnecting cable is over 100 ft long. (If you don't have any signal, check the coax!) Output of this amplifier, too, should be *at least* 1000 μV per channel, or more.

Your distribution system should not introduce too much loss. There should be a *minimum* of 500 μV at the antenna terminals of each set.

Somewhere in there you've got a weak tube or tubes, or a faulty dc power supply, or something like that. The signal levels will tell you where.

"Jig Smear" In Monitor TV

We have a small TV receiver that we're using as a monitor. We've remoted the picture tube from the cabinet, and housed it about 36 inches away. The picture lost deatil and seems to have a horizontal smear. Controls are working, but the picture just isn't as crisp as we want. What can we do about this?—S. L., Barrie, Ont.

This is very apt to be what I have christened *jig smear*. It is caused by the added shunt capacitance of the extension cables in the video circuits. You see the same thing when operating a color TV chassis on a test jig. It reduces the high-frequency response of the picture tube by shunting these signals to ground, etc. (You can run a video sweep test and make sure.)

Possible cure would be to isolate the pix-tube cathode lead from all the rest by a couple of inches. Try using a plain wire lead instead of coax or use very low-capacitance coax.

Fig. 7-6. Horizontal sync loss in VTR output could be due to bad head.

Videotape Sync Problems

I have an odd horizontal problem. It shows up as an offsetting of parts of the picture, making vertical lines look jagged. It's intermittent, and often comes out when the station changes cameras or programs. I can't do anything to the set that will make it show up. Seems to show up mostly on local programs.—W, L., Brooklyn, NY.

From the description and the symptoms, this could be videotape trouble at the station. It is either a complete loss of horizontal sync, as in Fig. 7-6, or a partial loss as in Fig. 7-7. Notice that the "sawteeth" or bends here are about 16 lines apart. One bad head on a standard videotape recorder will do this.

In color tape programs, this usually shows up as orange streaks (still 16 lines apart) across solid-color backgrounds.

Fig. 7-7. When VTR head deteriorates, horizontal sync stability suffers.

Chapter 8 Test Equipment

In this chapter, I'm going to mix apples and oranges, so to speak. The *text* is devoted to ways by which your existing test instruments can be "extended" to allow their application to measurements once reserved for more sophisticated (and expensive) test equipment. The *questions* at the end of the chapter deal with test equipment, too; but the questions are not from readers who want to extend the capability of their gear—rather, they represent problems some of our readers have had with their equipment. Just like a modern color set, an oscilloscope or a high-impedance voltmeter can go "on the fritz" too.

But first, let's see what our "conventional" instruments can do in the way of making "unconventional" tests.

CAPACITANCE

You can do a lot of things with "standard" test equipment. Use your imagination. For example, here are some typical questions, culled from the mailbag:

"*How can I measure the value of a 4000 μF electrolytic, when my capacitor checker only goes to 2000 μF?*" Disconnect it, and connect another one in series with it, watching polarity. If the other one is also 4000 μF, the combination is now a 2000 μF unit and you can read it. If the extra one is, say, 1000 μF the resultant capacitance will be around 800 μF. Capacitors in series, formula: the reciprocal of the sum of reciprocals. In plain English, this means dividing 1.0 by each value in series, then adding the resultant values, and dividing 1.0 by the sum, or:

$$\frac{1}{\frac{1}{C1} + \frac{1}{C2} + \frac{1}{C3} + \frac{1}{C4}}$$

By the way you can use the reverse of this to check a very small capacitor, if it comes out jammed right up to one end of the dial. Connect a known capacitor across the terminals of the capacitor checker (take the test leads off). Balance for the widest eye-opening or meter deflection, whichever yours does. Now connect the unknown across this one, and rebalance the bridge. You can usually read 3 to 4 pF capacitors by the increase in reading needed to restore balance.

CABLE BREAKS

I've got a long shielded audio cable and it's open! Is there an easy way to find out where the break is?"

Yep. Leave the capacitor checker turned on! Connect it right across one end of the cable, from inner conductor to shield. Balance the bridge and note the capacitance reading. This is the capacitance between the inner conductor and the shield. Actual reading isn't important.

Now go to the other end of the cable and repeat this test. Most cables break right at one end or the other, at the plugs. If you get a good-sized capacitance reading at one end, and practically nothing at the other, the very low-reading end is where the break is (Fig. 8-1).

If the two readings are almost the same, the break is somewhere near the center of the cable. For example, cable with 1.0 pF capacitance per foot, and 100 ft long: You read 60 pF at one end and 40 pF at the other. The break is 60 ft from the 60 pF end. (Notice how I make those figures come out even?) In actual cables it won't come out so even, of course, but the *percentage* will!

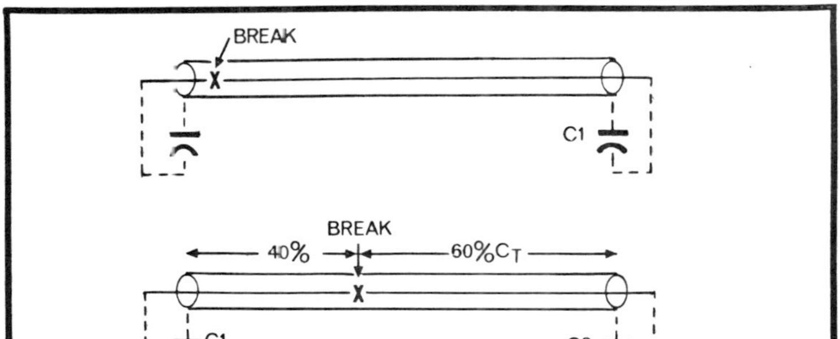

Fig. 8-1. To find a cable break, measure capacitance at each cable end. The value of your reading will tell you where to look: If both readings are about the same, the break will be in the middle. The lower the reading at one end, the closer the break to that end.

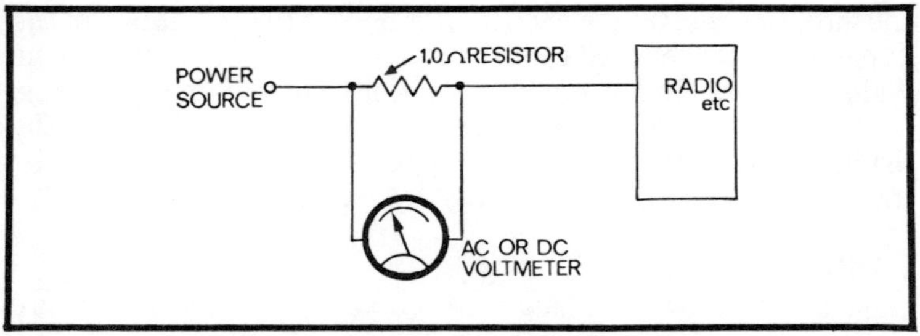

Fig. 8-2. A 1-ohm series resistor makes a handy-dandy ammeter. A voltmeter across the resistor allows "direct" monitoring of current, since voltage and current will be equal at 1 ohm (amperes = volts/1.0).

Take the two readings and add them. This is the total capacitance of the cable. Now measure the length, and you'll get the capacitance per foot. Do the mathematics and you'll find out where the break is from either end, in terms of the *percentage* of total length.

This is handy for checking out long rolls of shielded cable, coax, etc., on a reel, without unwinding the whole thing. In fact, you can even use this method between a good wire and a broken one—inside an intercom cable, etc. Read the capacitance between the known-good wire and the broken ends.

HOMEBREW AC AMMETER/WATTMETER

"They tell me that the circuit breaker pops at 1.25A ac; I haven't got an ac ammeter! How can I read it?"

Quite a long time ago, I "invented" an emergency wattmeter. I used this in a couple of books, and many articles. It consists of a 1-ohm resistor with an ac voltmeter hooked across it. Connect this in the primary of the ac power supply as shown in Fig. 8-2, and it will indicate 1 volt for each 1 ampere of current. Theoretically, this is an *automatic Ohm's law calculator*.

If you know the current, you can easily figure the wattage by multiplying by the line voltage. I tried it out on my bench on several different kinds of equipment and found it worked very well. For a check, I compared the readings against my pet wattmeter fed through an isolation transformer. It checked very closely.

The same gadget can be used to read milliamperes in high-voltage dc circuits. Just use a good 1000-ohm resistor. For every milliampere of current flowing through the resistor, you read 1.0V ac or dc; just be sure to use a high-impedance meter.

The same trick will work in low-voltage dc circuits, but you can't use a 1-ohm resistor—especially down around 6V or so. One ampere of current will cause a 1.0V drop across the resistor, and your set will have only 5V left! You can use this if you have a variable-output bench power supply. Just run the voltage up until you read the correct input voltage *at the set*. This will compensate for the metering-resistor drop, and everything will be normal.

There are some precautions you should be forewarned of. After publishing my little "Ohm's law calculator" trick, I got a letter from a reader, giving me fits because my "wattmeter" didn't work. He had used it in another one of my pet tests, for shorts in power transformers. This consists of disconnecting all loads from the transformer, then plugging it into a wattmeter. If the transformer is good, you'll see only a very small reading, about 3 to 4 watts. This is the iron loss of the transformer. If it has a shorted turn, you'll get a high reading, usually from 50 watts on up.

The irate reader claimed that he had tried this out on a large guitar-amplifier transformer, and it showed him a reading of 35 watts. So, he ordered a new transformer. Before installation, he checked this one, and it too showed him a 35-watt reading. (The problem turned out to be a leaky filter capacitor.) This is neither here nor there, in relation to this problem, of course.

So I rushed to the shop, got out a couple of old power transformers from the junkbox, and proceeded to run some tests. No-load current on the professional wattmeter showed the right answer. Practically no reading at all. Checking the same transformer on the resistor/voltmeter setup *did show a much higher reading*! On an old Philco TV power transformer with about 3 inches of iron, I got a reading of 0.62V. Figuring this out, it is the equivalent of 74.4 watts! Another power transformer showed a reading of 0.3V (36 watts).

Oh, oh! Something *is* wrong, but what? The answer finally dawned on me. The professional wattmeter is a dynamometer movement, or a "true wattmeter." It has two coils; a current coil and a voltage coil, plus a resistor. The resistor helps to correct for any reactance; so this type of instrument will read the actual power drawn, without regard for its power factor.

The resistor-voltmeter setup is apparently being upset by the fact that an *unloaded* transformer primary winding is acting as a more or less "pure inductance." So it causes a phase shift between voltage and current, and the resistor-voltmeter wattmeter is reading what someone called "wattless current." (Current out of phase with voltage.) This is causing the voltmeter to show a reading.

Further checks, and many, many tests in the past, show that a power transformer *under load* shows the same reading with either instrument. The loading of the primary brings the voltage and current back to an in-phase condition. Tests on pure resistive loads showed exactly the same readings on both meters. Incandescent lamps, soldering irons, heating elements, and so on: all matched.

So I came to the conclusion that this one test—the no-load transformer check, was the only one where the resistor-voltmeter setup was in error; or any test where the current through a pure inductance must be measured, of course. Actual testing seems to bear this out.

So, if you're using this test, with the resistor-voltmeter wattmeter, remember this. On an unloaded, *unshorted* power transformer primary, your reading will be high—higher than it ought to be, that is. If the transformer *is* shorted, you'll be able to tell, very easily. You'll get a full-scale reading of several volts, or many, many watts, a definite indication of trouble. Just for luck, you might check a couple of new power transformers and see what the reading is. As an average, this will be dependent on the thickness of the iron core. On the old, heavy power transformers, you'll probably see readings of 40 to 60 watts or more, due to the greater iron loss. On the typical power transformer, it'll run 30–40 watts.

There's a double-check you can always make on a suspected power transformer. Just plug it in, and let it sit there for about 10 minutes. If it's shorted, it will get very hot. A good transformer will be barely above room temperature.

HOMEBREW SCOPE "PRESCALER"

"How can I read the rf output of a little CB transmitter on my scope? I get nothing—just a line!" That's normal. The average service oscilloscope doesn't have much sensitivity at 27 MHz. So you're not going to get a heck of a lot of signal through your vertical amplifier. Connect a diode across the direct probe tip, and then modulate the CB transmitter with some kind of audio signal. For example, feed an audio oscillator through a series capacitor and resistor to the transmitter's mike input circuit. Then look for the tone-modulated frequency on the scope.

If you have a crystal-demodulator probe with your scope, use it. If the pattern isn't high enough, fasten about 12 to 15 inches of wire to the probe tip and couple it loosely to the transmitter's whip antenna. If you must make tuning adjustments, etc., tune for maximum pattern height. Incidentally, a No. 222 penlight bulb makes a good "dummy antenna" for very small CB transmitters. Won't glow very brightly, but will glow if the set is putting out rf. Solder the bulb to regular antenna plug used on set. Remove antenna.

There are lots of other ways to use your "common" test equipment, to make special tests. Use your ingenuity! You'll come up with lots of useful ideas.

THE VOM AROUND THE CAR

Here are a few handy tricks and tests you can make with a standard volt-ohmmeter (VOM) around a car, boat, motorhome or anything with a dc power supply system. The primary source of power here is a storage battery with a recharging system. These systems are about as simple as you'll get, electrically speaking. There will be a hot wire from the battery to all devices, and the frame of the vehicle itself serves as the return or ground. When the engine is running, the battery is being recharged by a generator or alternator. That's all she wrote.

Whenever you have any kind of electrical trouble, you'll need some way to get a handle on the problem. The first thing to check is the condition of the battery. Normal voltage of a fully charged battery should be 12.6V. Simple test for condition of charge: switch on a medium load, like the headlights. The voltage shouldn't drop more than a fraction of a volt. If it drops appreciably, the battery needs recharging.

For the simplest check on whether the recharging system is working, start the engine and then read the voltage across the battery. Of course, if the battery won't turn the engine over, that answers your question. If it does start, the voltage should go up to about 14.6V. Unfortunately, this won't tell you whether the battery will take a charge or not. The voltage regulation of the system is very good, and it won't let the voltage go higher than 14.6 volts, even on a battery with a bad cell.

You can start the engine with jumper cables from another car. There is a right and wrong way to do this, too. Watch the Polarity! All U. S. cars and a great many foreign models have a negative ground connection.

Connect the black cable from ground to ground on each car. Any good clean bolt, etc. makes a good contact. Now hook the red cable to the positive battery post of the good one, and to the positive battery post of the dead one. Start the engine of the helper car. Now try to start the engine of the dead one. This lets the generator of the helper car carry the load.

If you're using those chintzy dime-store cables made from aluminum, you'll probably be hard-pressed to get a decent connection between cars. Try this trick: Let the bumpers of both cars touch—solidly. Then connect *both* cables from positive to positive. This doubles the current carrying capability of your hookup, and will give an extra margin of power when the standard cable hookup won't quite cut it.

Cable Testing

If the car starts slowly, there may be other things wrong besides a low battery. Bad contact between the battery cable terminals and the battery posts is one. To check for this, put one prod of the meter on the end of the battery post itself, and the other on the cable terminal. Push hard to make a good contact, and then hit the starter. If you see any *voltage reading* at all, this contact is dirty. Use the lowest meter scale for this, for accuracy. If you see even 1V reading on this test, take the terminal off and clean both post and inside of the terminal. Bolt it back on very tightly. Recheck.

Second check: put one voltmeter prod on each post of the battery (not on the terminals; the post itself). Note the reading, then crank the car. The voltage should not drop below about 9 or 10V. If it does, the battery may be low.

If you don't get much voltage drop on the battery itself, take the same reading from the terminals. A drop here indicates a dirty contact between post and terminal; clean it up. If the voltage stays up at this point, but the starter cranks slowly, put one prod on the negative terminal of the battery and the other on a good clean bolt on the engine itself. Hit the starter; if you get any voltage reading, even on the lowest scale, the battery's ground cable isn't making good contact somewhere. You're reading a voltage drop across what should be practically a zero resistance. Loading it with the heavy current of the starter will show it up instantly.

Dim Lamps

In some cases, lamps on the car or motorhome will be mysteriously dim, though the battery checks good. This is usually due to corrosion or a dirty contact at the lamp socket. Take the bulb out, and scrape the terminals and shell clean. Check the socket for dirt or foreign matter.

If it's still dim, connect the negative voltmeter lead to the frame, and check the voltage on the hot wire at the lamp socket. If this reads the full normal voltage, take another reading, this time from the negative (shell) of the lamp socket to the nearest ground. This, of course, should be zero, if the ground connection is good. If you see a 2–3V reading here, connect a jumper wire from the socket to a ground. If this brings the lamp to normal brightness, take the original ground connection off and clean it. These are generally pigtails from the socket to a nearby bolt on the frame. Many lamps on modern cars are mounted in plastic; so they must have a pigtail to ground to make the lamp light.

If you get a low voltage under load at the lamp socket, take the lamp out and repeat the test. If you now get the full battery voltage, the chances are that you have a wire somewhere with almost all of the strands broken. The low loading of the meter will show you the full battery voltage, while the load current of the lamp causes a high voltage drop in the wire. Cure: Trace the wire, looking for a place where it has been almost cut in two or bent very sharply. In some cases, such as lights accessible from inside the trunk, you can run another wire from the bad socket to a good one; tape the new wire to the cable, and that's all you need.

Leaks

One of the most annoying things in the world is to go out on a cold morning and find your battery so dead that it won't even grunt. If there is a small current leakage anywhere in the system, it will drain the battery overnight. If you get it going, the next morning it could be dead again. This *can* be the battery itself; the only way to tell is with a "load-tester" at the service station. If the battery checks good, then you have a leakage.

If your multimeter has a DC AMPERES scale, preferably up to 10.0 amps, you can use it for the next test directly. If you have only a DC VOLTS scale, get a 1.0-ohm wirewound resistor, and connect the prods across this, on a low dc volts scale. You'll read 1.0V dc for each 1.0A of current flowing through the resistor. Now you're ready to find the leakage.

Take one of the battery terminals off the post (either one). Connect the meter (if it measures current) or the 1.0-ohm resistor *between* battery post and cable. Be sure that every normal load on the car is turned off, and the doors closed to turn off the dome lights. You should see zero current. If there is any leakage current, you'll read it on the meter. In my own car, some years ago, I had a steady reading of about 2.0A. This may not sound like too much, but it pulled a perfectly good battery down so far that it wouldn't even pull in the starter solenoid!

To locate the leaky device, just start disconnecting things, When you pull the wire off the guilty part, you will see the

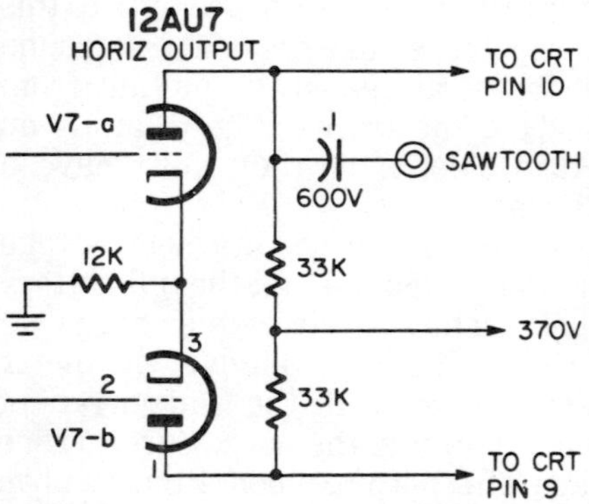

Fig. 8-3. Heath's IO-12 oscilloscope has jack output for sawtooth. Some models may not have this feature, but modification should be a simple matter.

leakage current disappear. In my car, it was the stoplight switch. A new one fixed it up. The push-on terminals used in so many places in modern cars make this a lot easier.

READER QUESTIONS

Heath 10-12 Scope

I need a source of sawtooth voltage for testing flybacks, etc. I have a Heath 10-12 scope, and I think I can get it from that; seems as if I read an article somewhere about it.

If so, where would I get it, and how?—J. L., Newark, NJ.

You may have an early-run 10-12 Heath scope. The schematic I have on this instrument *shows* a jack for a horizontal-sweep frequency sawtooth takeoff on the panel already (see Fig. 8-3). It is just a 0.1 μF disc connected from the plate of V7-a, pin 6, which goes to a jack on the panel. Strangely enough, I can't find any reference to it in the assembly instructions, nor does it show on the completed panel. It could be a modification.

There should be room for this jack on the front panel, say just above the HORIZONTAL INPUT, or between the EXT SYNC and 1V p-p CALIBRATING jacks at the bottom. This should give you all the sawtooth voltage you need. Be sure to use at least a 600V capacitor.

Calibration Drift

My RCA WO-33A scope has a problem! After adjusting the calibration, and using it for about 10 to 15 minutes, the vertical deflection drops to about half normal. Vertical gain control won't bring it back. If I turn it off and wait a while, it comes back—J. C., Palm Beach, FL.

You seem to have a really good thermal intermittent. Since changing tubes didn't help, this is apt to be a resistor that is drifting in value as it gets hot. Take a full set of dc voltage readings on the 6BR8 and 6BK7 tubes in the vertical amplifier, with the scope cool and working normally.

Wait until the problem shows up, then repeat the voltage readings. This should show up something. You can also use the old faithful heat-and-cool tests for thermal resistors. Heat them with a soldering iron, or spray coolant on them after the trouble shows up. Replace any that show a change in voltage.

Vertical Positioning On Scope

I'm having vertical positioning trouble on my scope. I can't get the line above the center of the screen. When it's first turned on, the line starts at the top and then slowly moves down the screen and won't come back. Where's the trouble?—G. P., Claremont, CA.

Vertical *positioning* of the trace on a scope screen is determined by the value of the *static dc voltages* applied to the deflection plates, usually "through" the final vertical amplifier tubes. (This accounts for finding the *vertical positioning* control in the *grid* circuit in some scopes.) The vertical output tubes are dc coupled to the deflection plates. Figure 8-4 is the basic schematic of this type of circuit. Here,

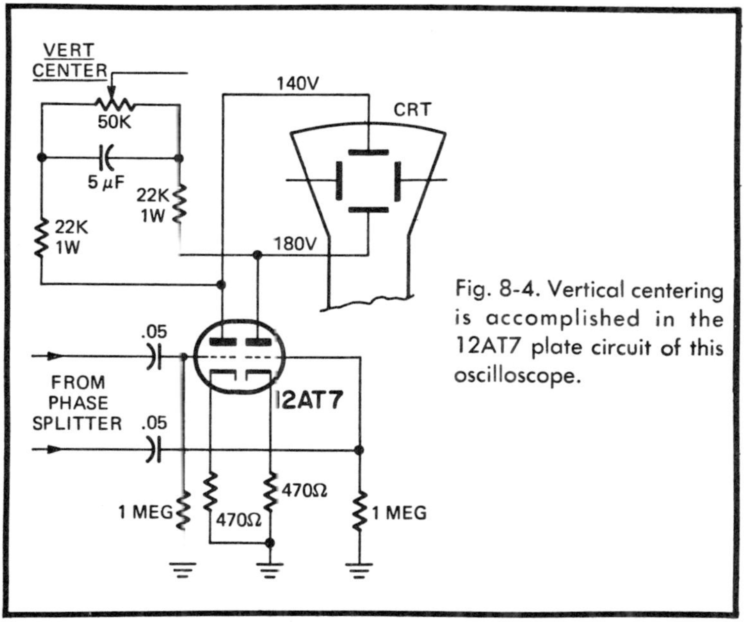

Fig. 8-4. Vertical centering is accomplished in the 12AT7 plate circuit of this oscilloscope.

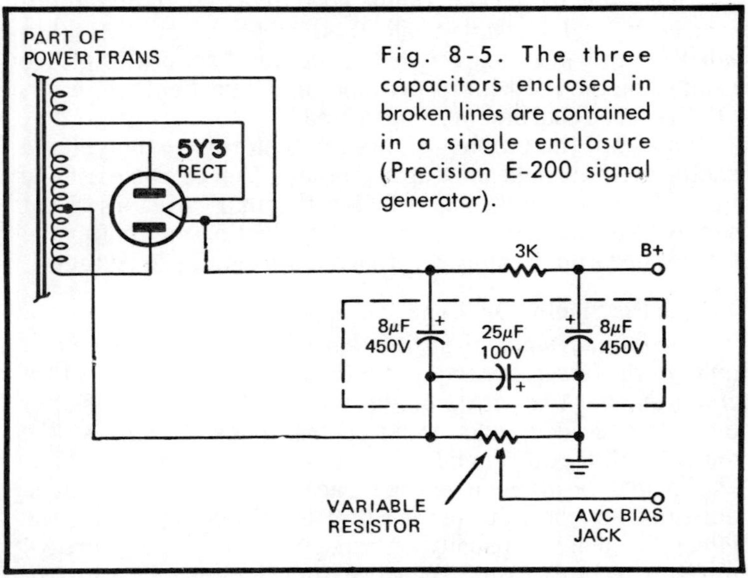

Fig. 8-5. The three capacitors enclosed in broken lines are contained in a single enclosure (Precision E-200 signal generator).

the vertical centering control is in the plate circuit. Dc voltages on the plates of the 12AT7 should balance. In your letter, you showed a good-sized unbalance here, and this is the trouble! Check tube and resistors.

Filter Capacitors

My old Precision E-200 signal generator has a shorted filter capacitor. I can't find the value of these on the can, and I don't have a service manual. What are they?—J. P., Lousiville, KY.

The main filters in this generator are 8 μF at 450V. There is also a 25 μF at 100V in the same can. Watch out for the connections on any replacements. There are actually four terminals (see Fig. 8-5).

You'll probably be better off using three separate capacitors; you might have a hard time finding a single-unit type with this exact hookup.

Reduce Full-Scale Reading of VTVM?

I've got an old but good VTVM, with a 5V full-scale as the lowest range. Do you know how I could modify this to get something like a 0.5 or 1.0-volt full-scale? A. K., Elmhurst, IL.

Sure. From Motorola, here's the solution: Connect one field-effect transistor and a couple of resistors in the circuit configuration shown in Fig. 8-6. This will give you a 0–1.5V

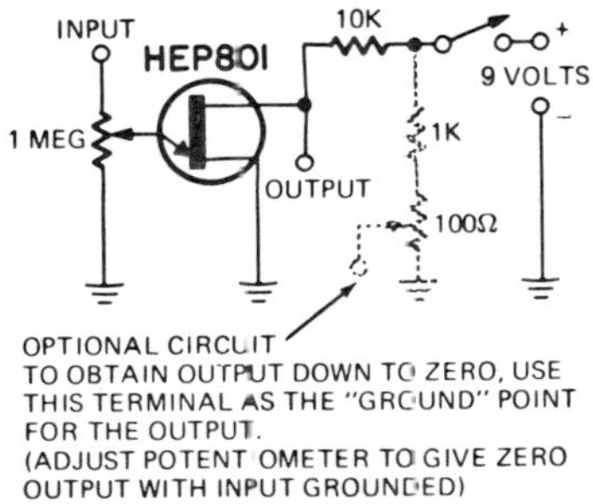

Fig. 8-6. This dc amplifier can extend the range of your scope of VTVM.

input for an output of from 0.4 to 9V. If you like, you can take it all the way to zero by incorporating the offset circuit (resistors in dotted lines).

VOM Testing

I blew the 1-ampere fuse in my VOM. After I replaced it, I can't get any readings on the dc volts, ac volts, or mA ranges. The ohmmeter works, though.—C. F., Tracy, CA.

If the ohmmeter still works you didn't blow the meter movement. That's good! Evidently the protective diodes saved it.

You can trace the paths from the input jacks to the meter. From the fact that several functions are out, you have an open circuit in one that is common to them all. Set the meter on the 2.5V dc scale and disconnect the meter itself. (To keep from slamming the 50 μA movement with the other ohmmeter battery.)

Now follow the circuit path from the + jack to the + terminal of the meter, as shown in Fig. 8-7. The "X's" indicate

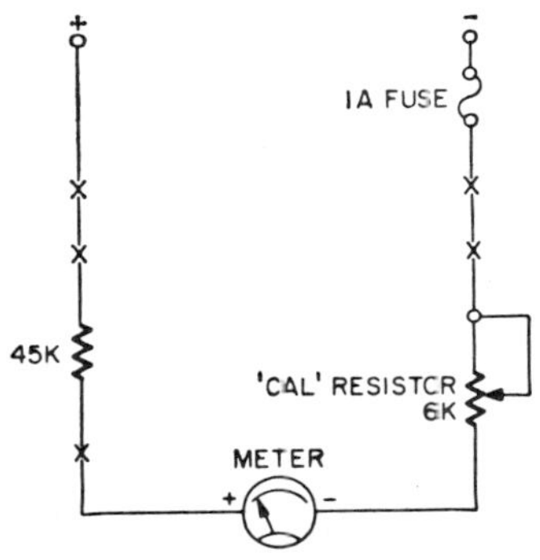

Fig. 8-7. The X's in this VOM circuit represent switch contacts.

switch contacts. If this path has continuity, follow the − circuit. It's somewhere in there. You'll probably see the charred remains of a precision resistor if you look closely.

The next most likely possibility is arcing at one of the switch contacts, when the surge that blew the 1-ampere fuse happened. This can make them open, from dirt or carbonization. In most cases, they can be cleaned up and made to work again.

VTVM Drift

My Heathkit IM-13 VTVM drifts. When it's on dc volts, the meter needle slowly swings to the right, all the way off scale. Acts up on the ac volts, also. The ohms scale seems to be okay.—E. N., Hensall, Ont.

The most likely cause for this would be an unbalanced "meter tube." This Heathkit uses the "meter between cathodes" circuit, a standard VTVM circuit. If the two triodes of the tube have unequal emission, or grid emission, you'll get this kind of trouble.

Age a new 12AU7 tube by letting it sit in the tube tester all day with only the filament voltage on it. Then try it. Recalibrate the meter on dc volts as per the instructions. You *might* have to try one or two new tubes, but I doubt it. One good tube will usually do it.

Shorted Scope Transformer

Mr. S. J. Martin of Camden NJ sends along this cute idea for fixing a short in a scope power transformer. He says: *The intensity control kept burning up on an old model Heathkit scope I bought as-is. Finally found that the pix-tube heater winding was leaking to ground. Everything else on the power transformer was okay. So I connected a standard TV "isolation" type brightener into this circuit, set on isolate, not boost. Worked fine!*

Thanks. I was going to suggest a separate 6V filament transformer, but this is probably easier to install and more compact. Good idea.

There's a Better Way

A while back you suggested a way to make shunts for old meters. If your shunt wire should come loose during the tests, the current could damage the meter. We use a safer method here.

The circuit (Fig. 8-8) is set up just as yours is, but the shunt wire is permanently connected at all times. One of the meter terminals is connected to one end of the shunt wire. The

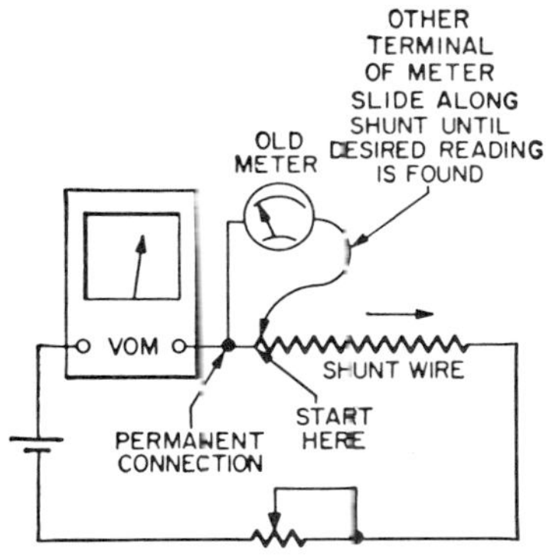

Fig. 8-8. Improved VOM shunt for reading high currents.

other meter terminal is simply slid along the shunt, starting at the "short" end, until the desired current reading is found.

The shunt wire is then marked at this point, the power turned off, and the shunt wire cut, wound up, and fastened tightly across the meter terminals.—F. S., Chicago IL.

Right! There's always someone around to work out a better way. Thanks!

5UP1 for 5DEP1 CRT?

I have a KG-635 Knight scope with a bad 5DEP1 crt. I've got a new 5UP1 crt; will it work?—F. B., Chicago, IL.

I believe so. The two tubes have exactly the same basing and, as nearly as I can tell, the same heater voltage and current. Knight used a 5UP1 tube in the KG-630 scope.

Scope Problems

In the horizontal sweep output amplifiers on my scope, one of the plate load resistors gets very warm, and the other doesn't. Tube okay. Dc voltages on the plates don't balance and they should. Trace looks funny too. Sometimes it goes completely out, then won't focus, then brightens up. Which part is causing this?—M. C. Baton Rouge, LA.

Would you make that "parts"? You have double trouble! The overheating plate resistor in that push-pull horizontal output stage is probably caused by a leaky coupling capacitor to the grid of that stage. Check the dc grid voltage with a VTVM. Too much positive bias will make that half of the tube draw too much plate current and overheat the load resistor.

Second trouble is very apt to be a drifting resistor in the network of resistors from the B+ supply to ground. Since these supply all of the voltages to the tube cathode, grid, focus

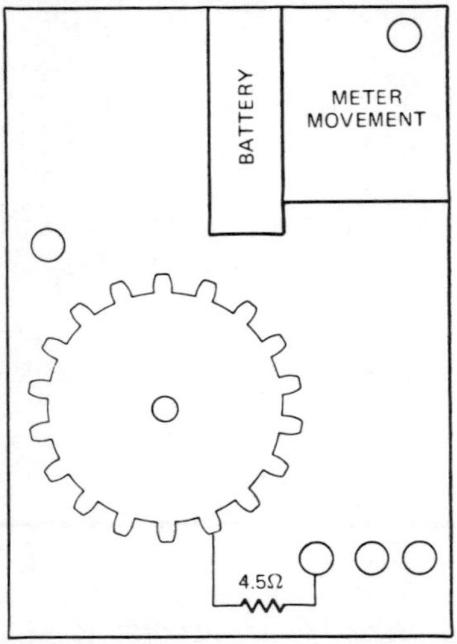

Fig. 8-9. Midland 23-101 voltohmmeter, back view (with cover removed).

electrode, etc., any change of resistance will upset the grid–cathode voltage and the brightness, or the focus, or all of them at once.

Always check for leaky coupling capacitors and resistors that have changed value. These cause practically all of the trouble. Fix them and your scope will be as good as new.

Burnt Resistor, Import VOM

I have a Midland VOM, Model 23-101. This is an import. There's a resistor burned up; goes from the "–" terminal to the switch. Can't read the value. Can you help?—H. E., Ottawa, Ont.

Yep. It is a 4.5-ohm wirewound (see Fig. 8-9). You can probably get a 1% resistor in that value or make one by winding fine resistance wire on a 2.2-megohm resistor. Use silk-covered wire.

Sweep Generator Output Low

My sweep generator works very well on i-f alignment, but I can't get enough scope pattern height on tuner alignment, especially on the high channels. What can I do about this?—E.N., Winnipeg, Man.

This is pretty much "normal." Most sweep generators use fundamentals on i-f's and even the low TV channels, but go to harmonics for the very high bands. When you double the frequency of an rf signal, you also *halve* the amplitude! Since we have a good-sized loss in our pickup probes, too, this often leaves us with not a heck of a lot of "curve" on tuner alignment.

You might do two things: One, go on through the first i-f stage and hook the scope to the second i-f grid. Shunt the tuned circuit in the first i-f plate with about a 10,000-ohm resistor, to flatten it and keep it from affecting the curve. Makes the first stage act as a "preamp" for the scope.

The other thing you might do—add a wideband rf amplifier between the sweep generator and tuner input. A good all-band TV antenna booster might work very well for this, since many of these have overall gains of up to 20–25 dB.

VTVM Problem

I've got a weird intermittent in my VTVM! On the ohms ranges, it can be zeroed, but if I switch to a different scale, the needle jumps around, and may come to rest off-zero. Same on dc voltages. Sometimes I read the correct voltages, then get only 1 or 2 volts! I noticed that once, when I pulled on the test leads, it changed. What's going on here—W. T., IL.

Offhand, one of two things. You may have an intermittent contact inside one of your test leads (which happened to me once, and caused problems!), or your grounding inside the meter may be poor! At times, the circuit board, internal framework, etc. may develop a poor ground connection to the front panel! This can really cause some problems. Take the meter apart and loosen the front panel. Clean the mating surfaces and tighten the bolts firmly.

No Vertical Deflection

I just got hold of an old but good Sylvania Model 132 scope. It's been pretty severely fiddled with, but I've got most of it working. However, I can't get any vertical deflection. Vertical amplifier stages seem to be okay. F. R., Flint, MI.

Check from the vertical deflection plate on the crt, which is pin 8, back through the wiring. Somewhere in there you'll find a switch marked VERTICAL DEFLECTION. It seems to be used to open the circuit from the vertical amplifier to the crt; for times when the signal must be fed directly to the vertical deflection plates.

It could be open or it could be defective. You can clean it, replace it, or if you don't need it, just jump it. This should give you vertical deflection. Sensitivity of this model, by the way, is 0.21V per inch.

Power Transformer

I need a replacement power transformer for a Knight KG-630 oscilloscope. Can't find anything on it.—R. W., Schenectady, NY.

I believe that a Thordarson-Meissner 22R 121 will work nicely in this instrument. It's not exactly the same size, but you should be able to mount it without too much trouble. Electrical specs are almost identical.

You can replace that original 6X4 rectifier with a pair of silicon diodes, and save that much heater current. This will also give you a bit more dc voltage output. This transformer has a 2.5V winding; this would be too much for the original 1V2 HV rectifier; try a stock color-TV focus rectifier in its place! Should work nicely.

AMPLITUDE-MODULATED MARKERS

I hear about sweep generators with modulated markers for setting traps. Can I do this with my Heathkit IG-14? Also, while aligning a CTC 7 RCA, I applied the negative bias to TEST POINT A *and lost my curve.—A. Z., Columbus, OH.*

Yes. For example, to get an AM modulated marker for the 41.25 MHz traps, punch the 41.25 MHz MARKER button and the MOD button at the same time. Feed the signal for the MARKER output into the i-f input, or tuner test point. Scope to the video detector, through the postmarker adder.

You'll see the 400 Hz modulation on the scope. Turn up the generator gain until you get a fair-sized pattern. Now, just tune the trap for *minimum* amplitude of the sine wave, and that's it. This can be done with any sweep generator having amplitude modulation of the markers.

You've just got *too much* negative bias if you lose the curve. Reduce this until you can see it again.

The Open Forum Chapter 9

This chapter is devoted exclusively to service problems as experienced by readers of *Radio-Electronic's* Service Clinic. The problems have been categorized according to general symptom, beginning with *voltage problems* (which ranks as the number-one source of problems for corresponding service technicians).

VOLTAGE PROBLEMS

I've got an odd problem with a Zenith 23XC36. Boost-boost rectifier was burned out, and the 6HS5 tube dead. Replaced these, and got a picture. No color, but good picture. Now, when I checked voltages, I found that I was reading 2200–2400 volts on the 1200V B++ line, and 1200–1300 volts on the raw boost terminal. Everything I've checked so far in this area is good. What causes this?—J. S , Alhambra, CA.

I know what I'd check first; my *voltmeter*. (That's not intended to be insulting; I've had the same thing happen to me.) Seriously speaking, if you really had voltages this high on the boost and boost-boost, your picture would *not* be good. It would probably be flaring so brightly that you couldn't even turn the brightness down enough. You'd have 2400V on the screens, for example.

It *is* quite possible to get readings like this in certain cases. If you are using a very sensitive FET VOM or VTVM with an *unshielded* "hot lead," the instrument can pick up *pulses*, rectify them, and then add this to the actual dc voltage on the test point. There is a very high-intensity field all around the area where you were taking the readings; this is normal around the base of the damper tube, yoke leads, and so on. I checked out a similar meter not too long ago, and got the same thing. By using a shielded lead to the hot probe, the problem was cleared up.

Try reading these voltages with a different voltmeter. I think things will look better.

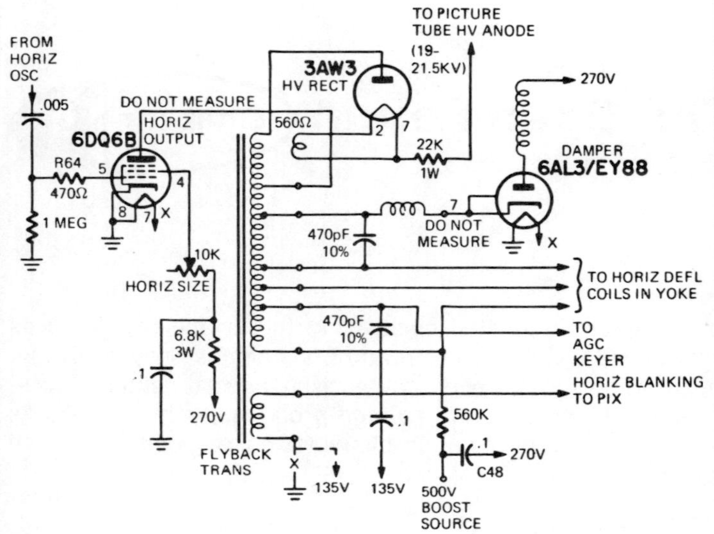

Fig. 9-1. Motorola TS-579-A B+ boost circuitry.

I replaced the high-voltage rectifier heater winding on this RCA CTC 25. Original was shorted. Now, I can't get the set to work properly! If I turn the brightness down to where I can barely see the picture, okay. If I try to turn it up, I get thin white streaks across the raster. I can get the same thing by turning up the pix tube screen controls. All tubes good.—G. S., Highland, CA.

It sounds as if the heater of the 3A3 isn't getting hot enough. This would account for the big drop in the high voltage when you increase the brightness (raise the pix tube beam current.) Recheck the new heater winding; be sure that you put on exactly the same number of turns around the core that the original had! If you're one short, your heater voltage will be low, and the cathode of the tube will be too cool.

While you're in there, look for signs of arcing across the 3A3 socket. Arcs, or a bad contact, can cause the "thin white lines" symptom. Check the 3A3 plate lead for corrosion, too.

This Motorola TS-579-A has the most weird group of symptoms I ever saw! Raster only half-width, high voltage low, boost low. However, the 6DQ6 cathode current is 110 mA; screen voltage okay, grid drive 150V peak to peak. Tubes good. (See schematic, Fig. 9-1.)

Suspecting the boost capacitor, I bridged it; no help. There's another one in there; bridged it. No help. Noticed that the schematic said the boost capacitor went to +135V. Thought this was a typo, and shunted a capacitor from boost to +270 volts. Whammo! I got the raster, boost and high-voltage back!

Later, I checked the Motorola schematic, and the boost capacitor does go to +135V! But, it won't work with it hooked there. What's going on?—J. G., Mena, AR.

That isn't a typo. The boost capacitor does go to +135V on this chassis. I thought the same thing, the first time I ran into it. After some intensive head-scratching, I tried the same thing you did—shunting the boost capacitor to the +270V. That other 0.1 μF capacitor you see in there is the boost *filter*; note that the boost goes through 560K before it gets there.

If you'll check that 10 μF electrolytic capacitor in the +135V line, you'll find that it's *open*! This puts the 12K resistor "in series" with the boost pulse, and raises the impedance so much that you don't develop any boost voltage. I found about 75 to 80V p-p of hash on the ones I had, indicating that the capacitor was wide open.

Without boost voltage (which is the *plate* voltage of the 6DQ6 horizontal output) you won't get enough output; you'll have half the high voltage and half the sweep width. This *is* rather unusual circuit, but the basic *reactions* are just the same.

With the switch in SERVICE *position, the high voltage is normal on this GE CA color chassis. In* NORMAL *position, it drops quite a bit, and I have vertical retrace lines in the picture. The horizontal scan also slowly collapses.*

Would you suspect the vertical output circuits? I replaced the vertical output tube; no help.—H. K., Chicago, IL.

Not the vertical circuitry, I'm afraid. This kind of symptom is usually due to one of two things: excessive beam-current in the picture tube, overloading the high-voltage circuits, or to a low-output high-voltage circuit (which would include such things as the high-voltage rectifier, horizontal output tube, etc.).

Check the grid and cathode voltages on the picture tube, in NORMAL position of the *service* switch. If the grids are too far positive, or the cathodes too far negative, the tube will draw too much beam current and overload the high-voltage supply. Maximum here is only 1500 *microamperes*. Check the setting of the picture tube screen controls, too. If these are too high, they can cause the same symptoms.

What you actually have is too much brightness! This will almost always let retrace lines show.

The 20V regulated supply on a Sylvania D-12 color chassis reads only 10V. I have no raster or sound. Higher dc voltages okay.—T. D., Bellevue, OH.

One of two things; either you have a high leakage or short in some of the many circuits fed by that 20V source or the regulator transistor itself is bad.

Disconnect the output of the 20V supply. Turn the set on. If this reads exactly 20V, it's okay. If it reads 10V no-load, try a new regulator transistor. Should hold the dc output constant at 20V, with line voltage all the way from 105 to 120 volts.

Loss of video and raster could be due to low supply voltage on the first and second video amplifier transistors, biasing the pix tube off.

I got this RCA CTC 24 in with a bad picture; weak, low brightness, etc. Finally found that the boost-boost voltage, instead of 1100V, was only about 850V, which is the same as the boost that feeds it.

I replaced the boost-boost rectifier with a new RCA part. Now, instead of having 850V on the boost-boost, I get only 300V! I can't find a short anywhere in the screen controls, which is about the only load on the thing.—H. L., Tucson, AZ.

Grasp your longnose pliers firmly in one hand and soldering iron in the other. Now take the new rectifier out, reverse it, and put it back. I think your boost-boost will come back! The newer RCA rectifiers used for this are not marked with polarity. However, the silver end of the thing is the cathode! (I will never tell anyone where I got this little gem of information, but believe me it's based on actual experience!)

I have no high voltage on this Zenith 14N23, 6JN6 cathode current is 180 mA, boost vootage +350V, drive waveform and voltage look okay. I can get a little bit of glow from a neon lamp near the plate of the high-voltage rectifier tube. I checked the yoke, and it seems okay. Any ideas?—L. C., Ink, AR.

After making all of these tests, there seems to be only one thing left. Check the flyback. You're getting a little bit of boost (not nearly enough), and your output tube cathode current is pretty high, indicating that the 6JN6 could be pumping into a short.

See if the HV winding of the flyback doesn't show a high resistance, or read shorted on a flyback tester.

I've got no high voltage on an Admiral H3 portable. Grid voltage on the pentode section of the 33GY7 is only about −8V. Screen grid voltage on the 8LT8 horizontal oscillator is okay, but plate voltage only about +10V. Control grid voltage on the 8LT8 is about +10V; should be −6V. I get positive voltages all over the diode section (afc) too. The horizontal oscillator isn't working, but why?—R. D., Reno, NV.

You have a leaky or shorted capacitor *somewhere* in the afc section of the oscillator circuit. Check C423, the 3900 pF comparison-pulse feedback capacitor from the 8LT8 output plate to the afc diode plate. This has caused exactly the same symptoms. (See Fig. 9-2.)

The dc leakage through this capacitor puts a positive bias on the oscillator grid and loads down the plate circuit so that it cannot oscillate. Nearly all the plate voltage is dropped across the 150,000- and 100,000-ohm resistors.

This Sylvania D-11 has no high voltage. I can't get the −40V on the grid of the 35LR6. The plate voltage of the 6CL8 horizontal regulator is off, too; only −0.6V instead of −40V. Checked resistors, capacitors; tried new tubes. R123 was off; this is 270 ohms, so I replaced it with a new one. Can you help me out?—J. V., Santa Clara, CA.

For the first step, take that 270-ohm resistor out (R123) and replace it with the right value. Someone has evidently changed it before. This is *not* a 270-ohm unit; it should be 47K,

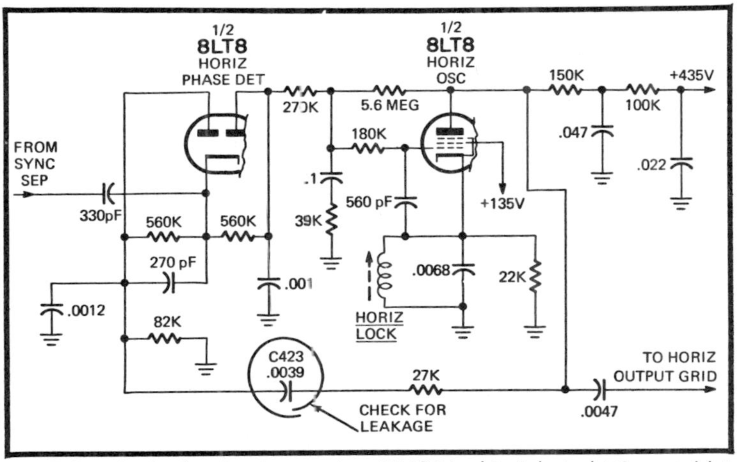

Fig. 9-2. A leaky feedback capacitor (C423 in this Admiral H3 portable) can cause a number of strange symptoms, not the least of which isvoltage loss.

and it's the plate-load resistor of the horizontal oscillator! No wonder you're not getting any grid drive on the 35LR6.

The plate voltage of the horizontal regulator tube (6CL8) is developed exactly like that of a keyed agc tube; from a pulse taken from the flyback. With no drive, no pulse; this should straighten out when you get the horizontal drive back.

This Sylvania 580-1 color TV is definitely a "hot chassis" type now, after being hit by lightning! Works on rabbit-ears, but kicks the breaker when the external antenna is connected. The den where the set's located has a tile floor, and the csstomer gets free shock treatments, which he doesn't enjoy!—C. I., Montgomery, AL.

Most likely cause; a short to core in the primary of the power transformer. It might even be a leakage of ac from one of the secondaries, since this set uses a half-wave doubler circuit, with a big electrolytic capacitor to chassis as the lower part of the doubler circuitry.

In any case, I'd check the primary first, then disconnect and check the secondaries for leakage to the core. New power transformer will probably clear it up.

I've got a weird condition in an Admiral IH2 portable. The fuse blew; I replaced it. Now, there's no high voltage, and the screen grid resistor of the 33GY7 gets very hot; smokes. I tried a new 33GY7, and got a raster for a moment, then it went out and the resistor smoked again.—D. Q., Omaha, NE.

Take a very close look at the *plate* connection of the 33GY7 socket on the circuit board; the plate connection of the pentode output section, that is. I believe you'll find that this will be pretty dark looking and probably have a very bad solder joint. Remelt it and add fresh solder.

Your output tube plate connection is probably opening. So you have no plate, and all of the current flows to the screen grid. There is only a 1-watt 560-ohm resistor here, and this promptly gets very hot. Double-check: read screen voltage at turn-on; if it jumps to normal, then drops to zero, this is it.

Hey; I'm in hot water again. This time it's an almost new RCA CTC 63XC with brightness problems. If I turn the brightness control or contrast control down, the raster will cut off. Won't come on again until I turn the set off and on. Everything I check seems normal; waveforms, etc. Help!—C. C., Baton Rouge, LA.

Stay in hot water long enough and you get pretty hard-boiled. Many late model sets have "protection circuits"

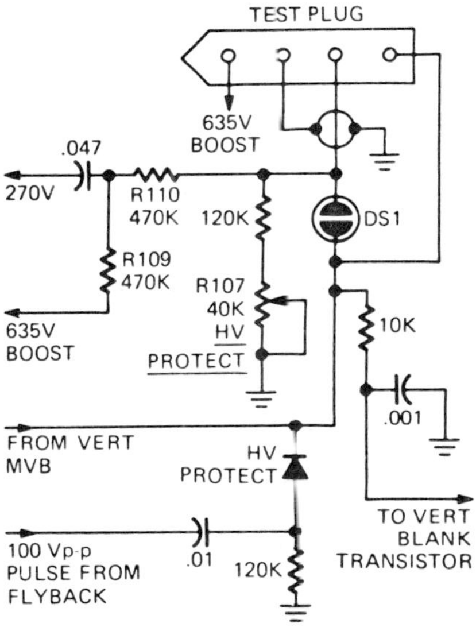

Fig. 9-3. RCA CTC 63XC high-voltage protection circuit. When this gem gets too eager to protect you, it will shut down the crt. Try adjusting R107 while varying brightness and contrast controls through their complete range.

built into them, to comply with some HEW regulation or other. This one does, too. The diagram (Fig. 9-3) shows the high-voltage circuitry. RCA's service data says "With a synced picture, connect a 5-megohm precision resistor across R109–R110. The neon bulb DS1 should fire, and produce a blank or no-raster." R107 is the control for this circuit, I'm almost certain. Seems to be able to set the threshold for the cutoff action (which cuts off the raster by biasing the video through the vertical blanker).

Check this: If the neon lamp *is* firing, the circuit is getting just a little too eager about protecting you! You should be able to adjust it with R107 to a point where it won't trigger on normal brightness or contrast changes. If not, the neon lamp may be defective; some of these will get erratic and fire at different voltages. Replace it.

I am having HV troubles with a Motorola TS-914. There is only 8 kV at the pix-tube anode. Screen voltage on 6JS6 is only 60V, and I read 34 mA current. Supply seems okay. Cathode current reads only 100 mA. I get −75V on the 6JS6 grid. I'm running this chassis out of the cabinet, with the yoke, etc. disconnected. Isn't that okay?—H. B., Utica, MI.

Nope. Not with this one. When you disconnect the horizontal yoke, you unload the flyback and kill the boost voltage. This *could* be the cause of your odd screen-grid reading. The screen is trying to carry some of what should be the plate current. Normal is only 18 mA. Also, in the TS-914, when you disconnect the *autodegaussing coil*, you open the *focus* circuit. Loss of the focus voltage will usually black out the raster, even with normal HV.

Hook it all up, using extension cables, and recheck. −75V on the 6JS6 control grid is too much. Should run about −60V. More than this will bias off the tube and reduce the output. There is a high-voltage control circuit in this chassis; diode fed by pulse from flyback, which regulates the bias on the 6JS6. Check this to see that it is okay.

If everything else seems to be okay, you might suspect the flyback. The HV secondary winding can be replaced without the other windings; it's made in two sections. HV winding could have a small short while the rest is okay.

The high voltage is down to 12V, I have practically no focus voltage, but the cathode current of the 6JS6 reads 220 mA. This is on a Heathkit GR-295 color TV. I am getting some odd readings on the grid-drive to the 6JE6, too. 200V p-p at the oscillator plate, only 150V p-p at the 6JE6; I lose all high voltage.—D. W., Vancouver, WA.

You do have a lot of odd symptoms, and they all seem to point to one thing; You may have a small short in the high-voltage winding of the flyback. Check all of the other things first, of course; boost capacitor, yoke, and so on. However, the high cathode current of the 6JE6, with all outputs so low, points toward an *overload* of some kind. This can be due to the short in the high-voltage winding, which will make the 6JE6 draw too much current. (Field feedback from thoughtful reader confirmed this: bad flyback.)

I've problems with a Zenith 25NC38. The high-voltage rectifier tube gets red hot and I can only read about 9−10 kV of high voltage. After some fiddling around, I pulled the cap off the 6BK4 high-voltage regulator; the high voltage shot up to 30 kV, and the high-voltage rectifier tube cooled off.

Voltage on grid and cathode of the 6BK4 reads +290V. Boost voltage reads +600V. I've checked everything I can think of in the 6BK4 circuit, no results. Where is it?—J. L., Elizabeth, NJ.

You're there, keep looking! The fact that you can get up to 30 kV with the regulator tube out of the circuit, and only 10 kV with it in proves one thing. That regulator tube is drawing a heck of a lot more current than it's supposed to. (It doesn't take much, either, with a plate voltage of 25,000 volts. Check the *maximum* rated current of a 6BK4 in the tube manual. It's a whopping big *1.5 mA!*)

You have the key voltage reading—you said that *both* grid and cathode of the 6BK4 were at +290V. If you read this from cathode to grid, you'd notice that the tube actually has *zero bias* on it! Same voltage on grid and cathode. Going back to the tube manual again, at zero bias the tube draws a fat 2.8 mA, or more than double the maximum rated plate current.

Normal bias on this tube must be about −15V. This holds the current down to not more than 1.0 mA. In actual operation, this is set by the two resistors in the grid circuit, 1.0 megohm variable (HV ADJUST) and the fixed 1.5 megohm to ground (Fig. 9-4). This goes back to the *boost* voltage through a 2-megohm resistor. The cathode is fixed at B+ voltage through the 1000-ohm resistor to the damper plate.

Normal operation should be at about −15V or less than 1.0 mA current with the pix tube screen dark. Check these resistors. One of them may be off; or C113, the 0.002 μF, may be shorted.

Im just getting started in TV, and boy, am I stuck on this one! I had no HV on this color set. I replaced the horizontal output tube and got absolutely nothing! I can read the full B+ supply voltage on the damper plate, but the damper cathode reads zero.—Anonymous, NY.

Check that damper tube! I think you'll find that the filament is good, but the cathode is completely open! Look at

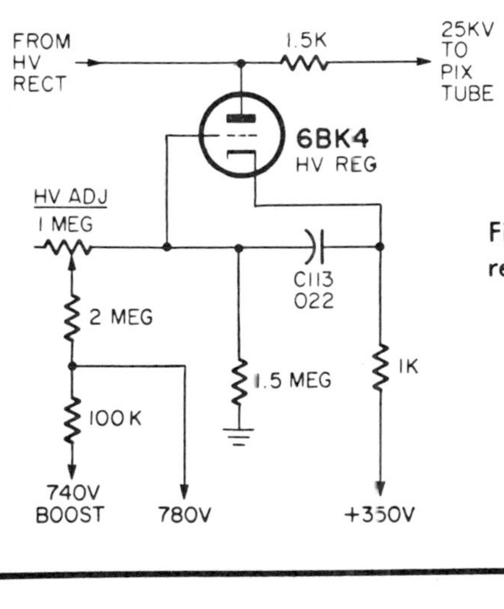

Fig. 9-4. Zenith 25NC38 regulator circuit.

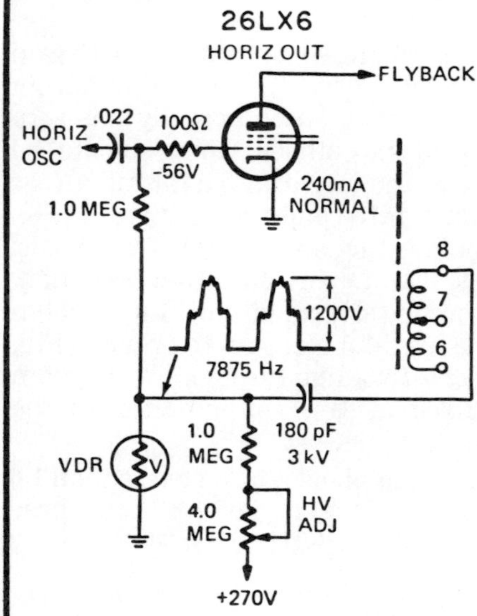

Fig. 9-5. In this Zenith 19CC19 chassis, regulation is accomplished by controlling the conduction of the horizontal output tube. Prime suspect would be the voltage-dependent resistor.

the base with a magnifying glass; you'll see the tiny metal "ribbon" between the base pin and the cathode sleeve broken. By the way, if you see "blobs" of metal on the ends of this, look out! This means that the cathode ribbon has been burned out by some kind of dead short in the damper cathode circuit.

I seem to have a high-voltage regulation problem in this Zenith 19CC19 color TV chassis. The raster blooms when the brightness control is turned, the high voltage fluctuates from about 16 kV to 27 kV at the same time, and everything is messed up in general. The HV ADJUST *control has some effect, but the regulation is very bad. I don't see any high-voltage regulator in this one.—F. D., Ames, OK.*

There is no high-voltage regulator, of the type we're used to, in this circuit. They use a "rectified-pulse" system that takes a 1200V pulse from the flyback and feeds it to a network consisting of a vdr, with a "bucking" (see Fig. 9-5) voltage from the +270V source. This controls the grid bias, and thus the conduction, of the horizontal output tube.

Check at the junction of the pulse-coupling capacitor, a 180 pF unit connected to No. 8 on the flyback, and the vdr. Be sure that the pulse is present at the correct amplitude. (By the way, there's a typo in the factory service manual. This pulse is shown as waveform No. 54, at 500V peak-to-peak. It should be the same shape as waveforms 19 and 20, but at 1200V peak-to-peak.

You could have one of two problems in this case. The 4.0-megohm high-voltage adjustment control could be open, or the vdr could be open. Disconnect the control and check it. If the bucking voltage is not present from B+, which acts to clamp this bias voltage, you'll get wild fluctuations. To check the vdr, lift the ground end and connect a milliammeter in series. Vary voltage (HV ADJUST) and see if you get about 1.0 to 1.5 mA current.

I've a TV set with a set of really odd symptoms. Everything seems to be okay, but no high voltage. I went off and left it running with the high-voltage meter hooked up, and in about 15 minutes it started ip. In about 2 hours it had gotten up to 5 kV. Later on, it came up a little faster.

Now, I can get a fair picture by running the ac line voltage up to 125V, but it still takes about 20 minutes to warm up.—L. F., Poughkeepsie, NY.

I'll tell you the exact truth; I haven't the faintest idea. However, you can get a good deal of information by reading the cathode current of the 6JE6. If it is very low at first, and gradually rises, this would indicate some problem in the dc voltages on the tube. A slow-heating damper tube could also do it; or an open (intermittent or thermal) boost capacitor.

In a Zenith 20Y1C50 color chassis, the HV rectifier fails frequently, and sometimes the fuse blows. I'm reading 33 kV on the HV at minimum brightness, and 30 kV at maximum. I suspect the HV regulator. Changed all tubes; no help. I've got +420V on the 6SH5 grid, 460V on the cathode, and 300 μA of current.—G. F., Lafayette, IN.

Your HV rectifier is being overloaded. When it fails it can blow a fuse. I've seen them short and blow not only a fuse but the horizontal output tube too!

You are not drawing enough current in the HV regulator. Your HV should be not more than 25 kV at minimum brightness. Normal voltages for this tube are +335V for the grid, +365V for the cathode, and a current of *600 μA*. This accounts for your high HV.

Check all of the capacitors and resistors around the 6HS5 HV regulator.

HORIZONTAL AND VERTICAL PROBLEMS

The horizontal output transistors and the flyback have burned out on a Sony 9-304UW TV set. They use two 2SC41 transistors in parallel. Can't find a replacement for these. Any ideas?—W. J., McL., Phoenix, AZ.

You *might* be able to replace these two with a single transistor with the proper ratings. For example, a Sylvania ECG-163. There are other similar types available.

For luck, run the ac line voltage up slowly, watching the collector current, output, etc. as you go. Normal current here is given as 520 mA per transistor. If the thing doesn't get too hot, fine.

Note: we got some "field feedback" from the reader on this. He tried the single transistor, and it worked. Got pretty warm, but worked. Just for luck, he replaced the surviving 2SC41 in parallel with it (one of them turned out to be good!) whereupon both transistors cooled off, drew about 360 mA apiece, and everything worked very nicely!

The horizontal oscillator in this Zenith 25MC33 chassis is very unstable. Horizontal hold range is very small. Blanking bar sometimes locks in the center of the screen. I hear a frying in the flyback, at times. Vertical sync very good.—J. B., Elmhurst, NY.

Short the afc. If this lets the horizontal oscillator run on frequency (that is, make a single, floating picture), the oscillator itself is okay. Since you have good vertical sync, this shouldn't be in the sync separator.

So the most likely cause here would be a defective dual-diode in the horizontal afc. Try a new one; this is the quickest way.

I've got a peculiar problem! One of my best customers brought in a little Emerson transistor portable TV, a 120771 chassis. It's got a "thermal shrink" of the raster, just like a weak horizontal output tube! The only trouble is, no tube! All solid-state!

This is the one which uses a GTO (gate-turn off) horizontal output device. About 15 minutes after it's turned on, the raster shrinks in about an inch from each side. I've read the current to the GTO, and it's quite a bit high. In other words, the thing is taking too much current, but the narrow raster shows that the yoke isn't getting enough current. I'm baffled! What the heck is this?—S. P., NY.

If your drive signal shows okay, and voltages are up to normal in the first few minutes of operation, I believe you'll find that this is due to a *leaky* gate-turnoff switch. Due to the circuit connection, this is effectively in shunt with the yoke. So you're trying to draw normal yoke current, plus the leakage current, and you wind up with a narrow raster.

The exact replacement for this is Emerson No. 815157. However, a good high-voltage, high-current silicon power transistor will replace it! Delco DS-502 or Motorola HEP-740 can be used as replacements; check the current to make sure; 680 mA is normal.

This little bitty TV set has a couple of big problems (Singer TV6). If I adjust for good vertical linearity, there's about 1–1.5 inches of picture chopped off, top and bottom. It also has a vertical sync problem that seems to get a little better as the thing warms up. Dc voltages throughout the vertical section close to normal. Oscillator and amplifier transistors replaced; no help. What do you have on this one?—M. S., Amarillo, TX.

This seems to have all the earmarks of an electrolytic-capacitor problem. Suspect any *coupling* capacitors in the vertical amplifier and output circuit. Low-voltage electrolytics can cause some weird symptoms. On the sync problem, check any *emitter* bypass capacitors in the sync amplifier, etc. If any of these are open, they'll reduce the sync amplitude.

The reader wrote back: *I found a thermal capacitor; the 10 μF coupling capacitor between the oscillator and vertical amplifier stage. When I cooled it, I lost the vertical sweep! Heat is up with a soldering iron, and it came back. What threw me was that this was a replacement* capacitor! Sync problem *was* due to an open emitter bypass, another 10 μF, on the oscillator transistor emitter!)

I have an odd problem in a Zenith 25MC36. It has a distinct horizontal jitter; picture good, horizontal sync good, color fine, and everything else, but it jitters. By accident, I discovered that if the sound was turned off, the jitter disappeared.—P. G., Mobile, AL.

Most likely cause, a loose *core* in the horizontal hold coil. I ran into one just like it quite a while ago. You can replace the coil, but you may be able to fix it by running the core all the way out, cutting a thin rubber band in two, and slipping it through the coil. Now, run the core back in, and the rubber band will hold it tightly.

There is no electrical connection with the sound circuits. The actual sound (air movement) is making the coil vibrate, changing the horizontal frequency very slightly when the core moves.

In a recent Service Clinic, you had a question about a horizontal oscillator running too high. I found the same kind of trouble in a Zenith 14M23. It whined, with no picture on the screen. Tracing the B+ lead back from pin 1 on the horizontal control tube, a 6GH8, I found that C44B, an 80 μF electrolytic capacitor in the B+ supply, was open. Replacing this cured the trouble.—Gerald Oium, Towner, ND.

Duly noted and filed! Thanks, Gerald. Found a lot of hash on the B+ when you scoped it, didn't you?

AFC PROBLEMS

I've got a fair picture on this Admiral 2H5 transistor TV chassis, but I get no snow at all with the antenna off. Weak-signal response very bad, and this set works on an antenna in a low signal area. My agc controls don't respond as they should. Where should I start?—R. G., Aurora, CO.

Run a step-by-step check on the *action* of the whole agc circuit. Start by verifying the keying pulse and video signals on the i-f agc keyer. Check the dc voltages on base and emitter; then take the antenna off and see if they *change*; they should!

This is athree-stage agc circuit i-f agc keyer, rf agc, and an rf agc delay transistor. The dc voltages on **all of these** should change when the signal is taken off or put on. If you find one that doesn't, there you are. (Watch out for bad solder joints! I found one of these with the same symptoms, and the emitter lead of the i-f agc transistor was loose.)

The color pops in and out in this Zenith 20X1C36. However, if I put the afc switch to off, the color stays in. I've changed all of the tubes in the color section; no help. What's wrong?—J. M., Phila. PA.

Nothing in the color section, apparently. This is an intermittent afc problem! Sounds as if it is detuned just enough to keep the color "hanging on the edge," and at any disturbance, such as noise, it lets go. Realign the afc, and it should stop the trouble.

I have replaced the flyback on a GE SB chassis, with a factory part. Now I get a split picture, blanking bar in center. I can unhook the pulse coupling capacitor to the afc, and make the picture hold normally; horizontal hold very sharp, of course.

The waveform from the pulse winding to afc and agc doesn't look right. It's more like a square wave than a sharp pulse. Could this be reversed?—W. M., Satellite Beach, FL.

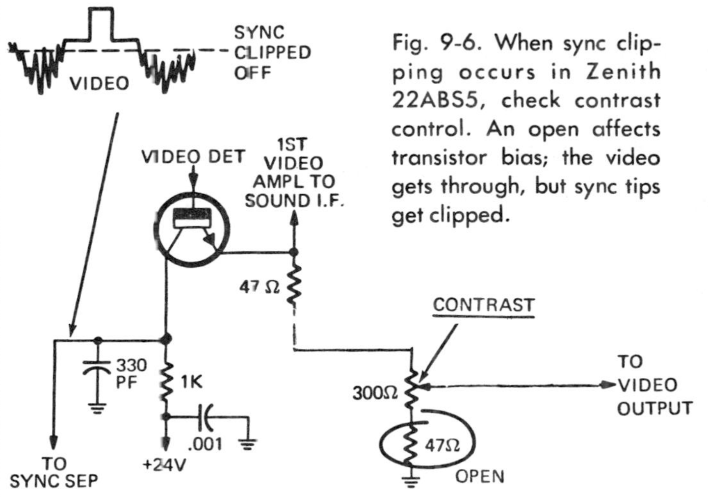

Fig. 9-6. When sync clipping occurs in Zenith 22ABS5, check contrast control. An open affects transistor bias; the video gets through, but sync tips get clipped.

It *is* possible that the pulse is reversed, of course, and this would account for the problem. However, since this flyback fits into holes on the PC board, the pulse winding (terminals 1 and 2) would have had to be reversed accidentally at the factory.

Suggestion; take both the agc and afc pulse-coupling capacitors off; and check the amplitude *and polarity* of the pulse, right at the flyback. Terminal 1 should be ground, and there should be a sharp 250V positive going pulse on 2.

SYNC PROBLEMS

In a Zenith 22ABS5 solid-state black & white TV, there is no sync at all. The picture can be stopped by very delicate adjustment of the holds. Video looks normal in amplitude. A scope shows the signal at the sync input with the sync clipped off like it had been shaved with a razor. Transistor good, by substitution. Collector voltage is high.—R. Q., FT. Smith, AR.

Check your contrast control (see Fig. 9-6). If it has no effect at all, the circuit is probably open. This upsets the loading on the emitter, and lets the transistor clip the sync, though the video goes through. I found the same thing in one of these, with a weird cause; the lower 47-ohm resistor was open, and it looked as if the wire lead to the resistor had *fatigued* and broken. Of course, I thought someone had accidentally clipped it, but examination with a glass showed it wasn't clipped. This upsets video to both sync and agc.

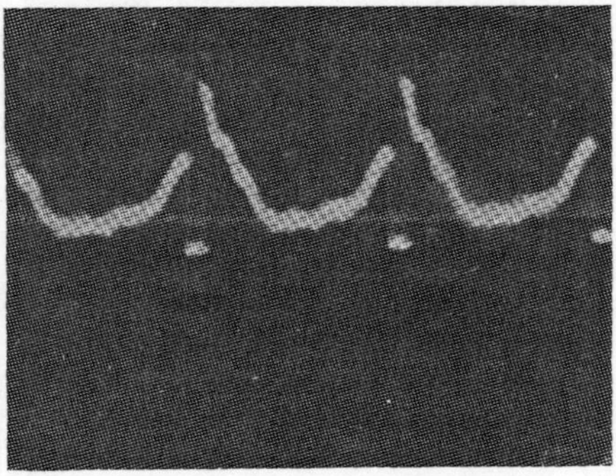

Fig. 9-7. A leaky afc diode causes a distorted waveform that should be a sawtooth.

I've checked and replaced every part in the horizontal afc in this RCA KCS-144—diodes, capacitors, resistors, etc. Still won't work. It will hold pretty well, until anything disturbs it; change of camera, commercial, etc., then out it goes. I'm at wit's end.—R. S. G., Provo, UT.

Last resort, on this type of problem. If it will hold *fairly well*, or hold at all, apparently without any sync, the oscillator is working, and the afc circuit itself is probably okay. One thing that has caused exactly this problem, in this and many others, is broken PC board conductors. These will be in the conductors which feed the horizontal sync from the sync-separator into the afc. I found one with a 1/4-inch break it, where a screwdriver had slipped at some time in the past. (Under a metal brace, so I couldn't see it, naturally.) Check all of these with an *ohmmeter*.

I've got a Zenith-14N22 chassis with very poor horizontal hold action. From the solder joints, it looks as if there has been some "work" done around that circuit. I've checked all of the capacitors, and so on, with no results. Where do I go from here?—F. S., Acorn, AR.

Check the afc diodes. If they have been replaced, it could be that someone used a poor grade of diode, or that one of them is leaking. The scope photo (Fig. 9-7) shows you what the sawtooth will look like when this happens (or it did on the last one of these that I found).

This waveform reads about 4 volts peak-to-peak, and it ought to be about 17 volts. Note the distortion; it also shimmered so badly that it was hard to photograph. Try a new pair of diodes, and check them for balance and reverse leakage before you put them in!

The complaint on this Philco 12N50A was loss of horizontal sync. I found that if I kept the contrast control below midposition it worked fine. When I turn it up, it works just like an agc control turned too far. A. E., Mascoutah, IL.

The contrast control in this set varies the gain of the video output tube by changing the screen grid voltage. I'd suggest checking the values of all resistors in that circuit. More likely, perhaps, that 5 μF electrolytic screen bypass. If it is open, you'd get a severe degeneration, and this could be causing what looks like a very severe sync clipping. The video signal for the sync separator and agc comes from taps in the video plate circuit. So if you're clipping, it would upset all of them.

I've got a sync problem in a Motorola TS-934A. Both vertical and horizontal sync are poor, and I don't have a heck of a lot of color sync either. Waveforms around the sync separator don't look too bad, but the collector voltage on the sync separator transistor is only about half of what it should be. Agc action is funny, too.—J. S., Tyrone, NM.

You may have double trouble here. First, check out that sync separator stage. The collector voltage on this transistor is *developed* across the emitter resistor of the horizontal sync pulse shaper transistor. So, if this voltage is low, the sync separator isn't drawing normal *current*. This could mean incorrect bias or a leaky transistor.

No. 2: Note that you have a *set of symptoms* which could point to some trouble in the *video* signal, as well as their dc voltages, this could be a "common cause."

Check the peak-to-peak amplitude of the video signals on each of these stages. Then, if it isn't high enough, follow it back to the takeoff point. This is the collector of the chroma sync amplifier stage on the SA panel. If your *picture* is okay, the video stages themselves should be okay; then, there's only one stage between this and the sync/agc circuitry on the FA panel. (Check the No.9 pin on the SA panel, and the interconnecting wiring to pin 6 of the FA panel, too.)

They brought us a Motorola TS-597 chassis that had been dropped by movers. We replaced a couple of tubes, straightened up some more, and got pix and sound. Now, all we need is some horizontal sync. The picture will barely hold, but if you move the core of the stabilizer coil even a tiny bit, out goes the sync. Very odd.—H. B., Hopewell, VA.

This has a very familiar sound! If the picture will hold at all, the chances are your hold coil, afc diodes, etc. are okay. However, the reactions you're getting are those of a complete loss of the horizontal sync to the afc stage.

Under the circumstances, I'd start at the plate of the sync separator and check for correct amplitude horizontal sync at that point. Now, follow this through the PC board to the afc unit. You'll probably find a hairline crack keeping it from getting to the diodes.

I've got an odd one on a Lafayette RK890A 8-track cartridge recorder. Playback and recording on the first 3 tracks, fine. However, when I record on the 4th track, the left channel is normal but the right channel is very low. The VU meter shows a low level on the bad one, while the other is okay. For a final weirdo, prerecorded tapes play normally on the last track.—C. C., Bronx, NY.

To change tracks, you *move the heads* up and down, right? There is no electrical switching as far as the signal is concerned. Track 4 would probably be the one with the heads in the lowest position, or closest to the chassis.

From all of the symptoms, and especially the reaction of the *VU meter* (which is right across the head, in RECORD, and works normally on all of the other three channels) it looks very much as if you have a *short* to chassis when the head is in Ch. 4 position. The very small wires to the head may be rubbing on a sharp edge, or something.

Get a big magnifying glass and a strong light, and see if you can tell where this is taking place. You might try putting plastic tape across any sharp metal edges near the wiring to the heads. Might even be able to verify this crystal-ball diagnosis with an ohmmeter on the head, moving it into Ch. 4 position manually.

I've got two little Motorola GP20GW cassette tape recorders with the same problem. They play back perfectly, but they record with a terrific distortion. The sound comes through, but there's a low-frequency hum or buzz recorded with it! I've tried two or three different mikes, and tried to check the bias, but no luck. What is this?—J. G., Mena, AR.

Keep on trying mikes! You haven't got the right one yet! Suggestion: get the original microphone, and try it. If you try to use this particular machine with a very high-impedance mike, such as a ceramic, crystal, etc., it will oscillate! This is causing the distortion. (No bias on this unit; it uses dc bias for record and erase!) Original mike is a low-impedance dynamic.

I have a Panasonic 8-track auto tape player. Works fine in the car, but when I hooked it up to a homemade 12V dc power supply for use in the house, the pilot light flickers with the loudness of the music. It doesn't do this in the car.—F.V., San Fransisco, CA.

You've got a little voltage regulation problem. The flicker is due to voltage drop on loud passages in the music. In all class B transistor amplifiers, there is a great difference between the resting current (no signal) and full-volume current.

Your power supply will give you the right voltage under small load, but on loud passages the supply voltage is dropping, causing the flicker. Try using a bigger output filter capacitor; it may help. If not, you'll have to get a power supply with a greater *current* rating.

I've got to install an intercom system for a lumber company. They want two speakers in large warehouses, some distance from the office. The system is an old one, with a 35W4, 12AU6, and 50B5. I tried this in one of the warehouses. I got very little volume and a lot of hum on "talk." What can I do to match the speakers better and get more volume?—G. R., Greensboro, NC.

Without a lot of trouble, not very much! This system does not have much gain, and only about 1 watt of audio output. Your best bet would be to use two more master stations in the warehouses, instead of the normal remote units. This will get rid of the hum problem.

With a brand new battery, this Motorola 8X26E transistor radio works perfectly. However, as soon as the battery ages just a little (internal resistance rises) the thing motorboats. If I bridge a 10 μF capacitor across the battery, it works fine. Battery voltage doesn't have to drop; it acts up with the full 6V applied! What should I do?—M. B., Spartanburg, SC.

Leave that 100 μF capacitor in there. There is no battery bypass in this set. "Motorboating" is a low-frequency oscillation that is practically always caused by feedback through the dc power supply. A 100 μF capacitor at 25V shouldn't be too big to tuck away somewhere in the case.

This 10-year-old Knight tape recorder is okay on playback, both channels. On record, the right channel is good, but the left channel is low and garbled. Tubes okay, dc voltages seem normal.—C. M. Whittier, CA.

Good playback on both channels means that the tape heads are very apt to be okay. Good record on the one channel means that the bias oscillator *is* working. Low output and garbled

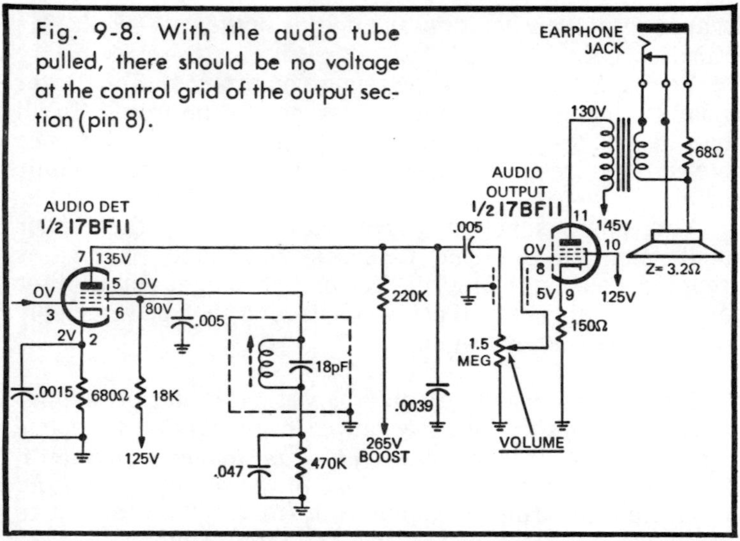

Fig. 9-8. With the audio tube pulled, there should be no voltage at the control grid of the output section (pin 8).

sound on only one channel looks very much as if you're not getting the *bias* to that head.

This would most likely be some accidental short or open in the wiring or in the switching to that head. Follow the signal through to the head with a scope and you'll see it.

I've got a GE portable, with a funny sound problem. The longer it's on, the worse it gets. I found that the control grid of the 17BF11 audio output section was going positive, up around +15V. I changed the coupling capacitor, but this didn't help.—V. S., Dalton, PA.

Pull the 17BF11 tube, and turn the set on. Of course, you'll get no raster since the heater string will be open. However, you'll have B+ on the plate (pin 7, Fig. 9-8) of the audio detector. Take a reading at the grid terminal of the output section, pin 8. This should read zero. If there's any leakage across the socket, etc., it will show up here.

If this test shows no leakage, replace the 17BF11 tube. The chances are that it has developed internal leakage, like the old 50B5s, etc. used to.

I've got a Zenith 10YT26 stereo amplifier, and it's weird. The customer jumped the fuse after it blew, and several resistors were burnt up. Still blows the fuse. Draws a very heavy current, and I can't find out why. I can get about +15V on the collectors of the output transistors for a moment.—J. D., Greenville, OH.

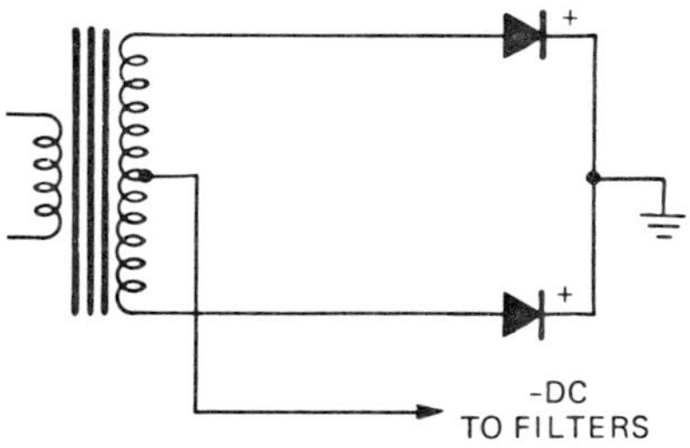

Fig. 9-9. This is the correct configuration for the power supply diodes in the Zenith 10YT26 stereo amplifier. Reversing the diodes delivers positive rather than negative B+ voltage.

No wonder. The customer evidently did something else beside put a 30-amp fuse in it. If you're getting a *positive* voltage on the output collectors, your rectifiers are *reversed*! The diagram (Fig. 9-9), shows how these should be connected. The positive voltage on the filter electrolytic is causing them to draw a heavy reverse current; incidentally, it would probably be a good idea to replace them too. This can damage even a good capacitor.

I need a 6AD7 tube for an old radio, and I can't even find it on the lists. Do you know of another type that will replace this? I'll change the socket if I have to.—T. W., Cupertino, CA.

You'll have to! There's no direct replacement for the 6AD7 triode/pentode. However, a type 6BM8 looks as if it has almost the same characteristics. You'll have to use a 9-pin miniature socket. Connect just as the original was, plate to plate, etc.

I need a replacement transformer for a Magnavox CR-703 radio. Can't get one anywhere. Can you help?—D. G., Los Angeles. CA.

Yep. Your original i-f transformer has a tertiary winding. This is switched by the "*AM broad/AM sharp*" selector switch, to provide a wider bandpass. To replace the transformer, you'll have to give up this feature.

A stock J. W. Miller 455-kHz i-f transformer, No. 14-C1, is the proper replacement. It is an exact duplicate as far as mounting is concerned. Ignore the tertiary winding, and hook it up as a straight i-f and it will work nicely.

I have no pedal tones on a Hammond Model M organ. Others work fine, but no sound from the foot pedals. What could cause this?—L. N., NJ.

In the early L series, this was usually due to a bad 6BA96 pedal amplifier tube. This is on the amplifier chassis, between the rectifier tube and the 6V6s. In the M series, it's in the same place.

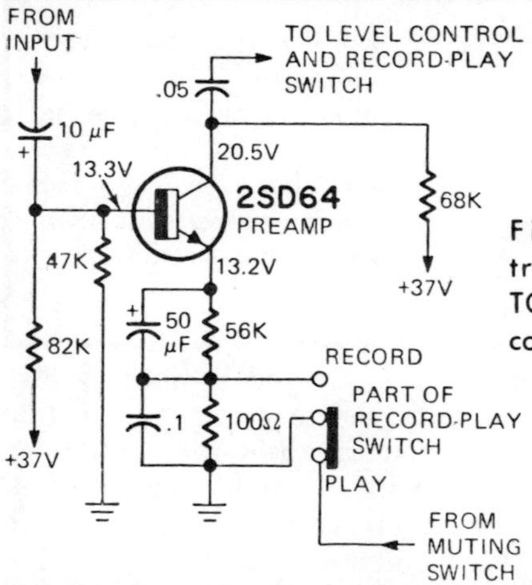

Fig. 9-10. Preamp transistor stage, Sony TC-102 hybrid tape recorder.

There are also a couple of added controls on the M. Check the pedal cutoff control and the pedal keying contact. Also, the flexible wire to the pedal drawbar above the upper manual. (Don't forget to check the 6AB6 tube and its voltages, too.)

I can't get anything through the amplifier of this Sony TC-102 hybrid tape recorder. The dc voltages on the 6AU6 second of amplifier are about +60V, plate and screen, but I read +8V on all three elements of the preamp transistor. That doesn't sound right.—H. C., Monroe, LA.

It certainly doesn't. From a reading like this, you have one or two possible things: the transistor is shorted to all three elements, or there is a resistor open in the supply! Take the transistor out and check it. If it shows a normal diode effect reading out of circuit, then turn the power on, and check the voltages on the transistor connections on the board.

From these readings. I'd suspect something like an open collector load resistor (R9, Fig. 9-10).

I've just replaced all four output transistors in a Truetone 4DC5929A-86 stereo amplifier. The top transistors in each channel had gone bad. Now, the right channel works nicely, but the left channel is distorted. New transistors were taken out and rechecked, but they were good.

I notice that at "no volume" (no input signal, volume off), the left channel output transistors get warm and show too much bias voltage.—R. T., Tulsa, OK.

Check the *driver* transistor. This is directly coupled to the base of the upper output transistor. So the collector current of this one sets the base bias of the larger output transistor. If this transistor has changed its characteristics, or if it doesn't exactly match the new output transistor, the bias will be off (just as you found it). This is a complementary-symmetry output stage (Fig. 9-11). So the resting current should be pretty low. Or, at the very least, it should *match* that of the other channel.

Take the driver transistor out and check it; better still, replace it.

The replacement transistors for this output stage could be RCA SK 3024/2035. A good replacement for the driver transistor could be another SK-3025. After replacing the driver, recheck the voltages and the balance between R and L channels.

The AM works fine on this Zenith 9ZT15 radio, but I can't get anything on FM. It's completely dead. Where would you look for this?—E. B., Adams, MA.

If you can get an AM signal through, the first two i-f transistors and i-f transformers are okay. They work on both AM and FM. However, there is a 3rd i-f transistor which is used only on FM. Check this one. It's probably open or shorted. For a definite check, feed a 10.7 MHz AM signal into the i-f input. If this won't go through, that's it.

If it does, the trouble is in the FM oscillator or rf transistor, or the FM tuner; loose solder joint, coil, etc.

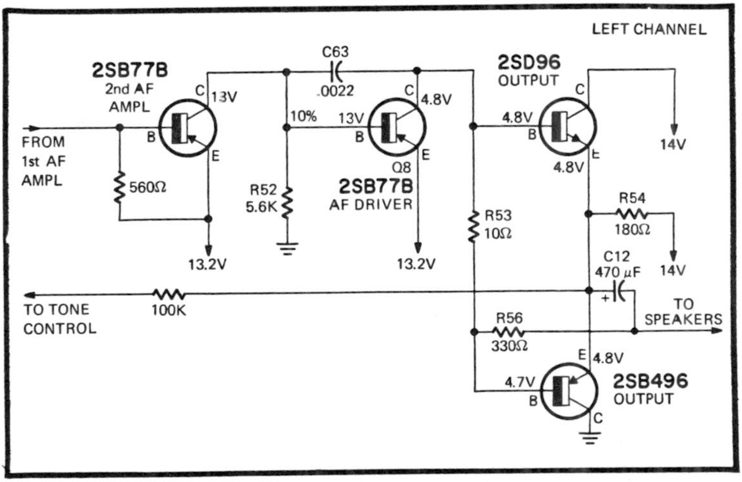

Fig. 9-11. Truetone 4DC5929A-86 stereo amplifier uses complementary-symmetry output stage.

This Plymouth 9BBCC radio will only play on strong local stations. When I opened it up, I found that someone had connected a 0.001 µF capacitor from the antenna trimmer to the emitter of the rf amplifier. This capacitor should go to one of the tuner coils; when I put it back there, nothing. What's going on?—J. S., Decaturville, TN.

Nothing is "what's going on." As far as rf gain is concerned, anyhow. You'll probably find that the rf coil is open, and that someone found out that he could shunt the signal around it, to the rf stage. This practically kills all gain in the rf stage, of course. Replace the coil, and realign.

I'm an old tube-type technician, and I do get lost in some of the new solid-state stuff. Take this Sherwood S-7100A. While it's cold, it will play in the stereo mode, then it will switch back to mono. Plays fine, but not stereo. Need advice.—E. S., Vancouver, B. C.

If you do not lose the *signal*, but only the stereo effect, this is a loss of pilot carrier. Signal-trace this through the multiplex decoder circuit, frequency doubler, and so on. You'll find a "thermal" part somewhere along the way. Your scope will show this up very easily.

I'd like to make a wireless remote control for a PA system. This is one of those "console-on-casters" types, which makes it hard to follow around! Who makes such items?—A. A., Lansing, MI.

Right now, I don't know of anyone making a wireless remote control specifically for PA systems. (Someone will write me an indignant letter inside of 3 days and tell me they've been building 'em for 30 years.)

I believe your best bet would be to locate a junk TV set with an ultrasonic remote control system; or if you're feeling flush, go and buy the remote control system from the distributor! There are quite a few makes which have very reliable ultrasonic remote control systems, with compact, hand-held transmitters, and transistorized receivers. These have volume controls "already built-in" and it would be a simple job to convert one for PA work.

I've an old Telechron electric clock, a 4H12. It ran slow for a while, then stopped entirely. My wife is very fond if it (it's her kitchen timer!) and she wants me to fix it. I think the drive coil is burned; it's so old that I don't know where to get one. Can you help?—D. D., Ink, AR.

I have the same kind of problems. (Don't we all?) Anyhow, try this: check the resistance of that coil. If it's around 600 ohms, it's probably okay. Turn it on and hold a small screwdriver near the frame. If it buzzes, the coil is okay. Your gearbox is probably gummed up.

Take out the 4 screws; the frame will come apart. Try some very light oil on the tiny pinion. If it won't go in, then put the unit over a soldering iron until it's pretty hot. You should see tiny bubbles at the pinion. Now, take it off, and apply the light oil to the pinion: at the same time, spray coolant on the case! This will create a little vacuum inside the case and suck the oil in. If standard light oil won't work, try using some of the silicone lubricant in spray cans. This is very light, and will get into places where oil won't.

This Craig 2108 recorder will play prerecorded tapes very well. On record, it garbles so badly that you can't use it. It won't erase, either. What causes this?—J. B., Taylorville, IL.

In this machine, they use dc bias for the erase head. The switch of this also grounds the emitter of the record-bias oscillator. Since the playback head is also used for recording, it would seem that this is okay: also the amplifier.

Check that part of the PLAY-RECORD switch. You can tell whether it is working properly or not by reading the dc voltage on the bias-oscillator transistor emitter (see Fig. 9-12). This should go down to −0.6V in RECORD if the switch is working properly. Check for bent contacts or broken lugs or wires.

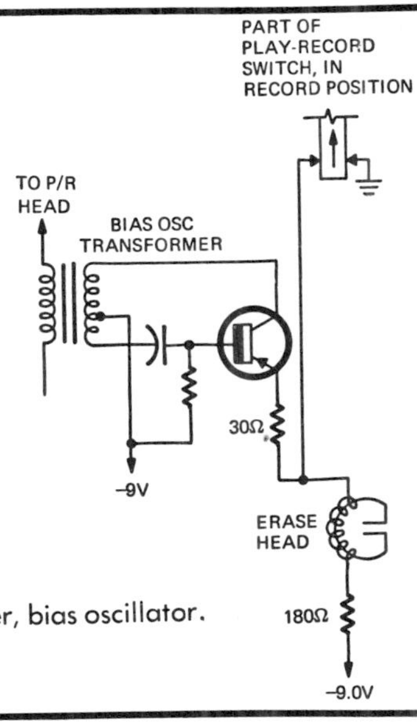

Fig. 9-12. Craig 2108 recorder, bias oscillator.

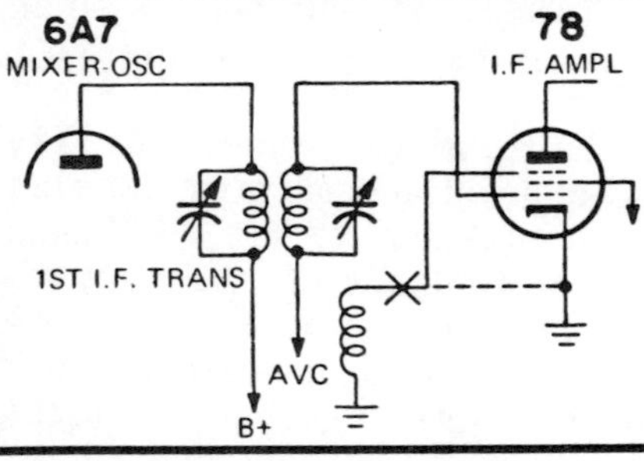

Fig. 9-13. I-f transformer in Philco 39-45 was a 5-lead type; it can be replaced with a J.W. Miller 512-C1.

One of my customers has a Silvertone radio which he plays in his barn. The 50HK6 output tube burns out about every two months. In the shop, everything is normal; no shorts, etc. What could cause this?—R. B., Springfield, OH.

The first thing I'd check would be the ac line voltage in the barn. In some cases, rural lines run higher than they should, up to 135V in some cases. This would shorten tube life.

If this is it, you can add a small line-ballast resistor to hold the voltage down to about 105V for best tube life. Check this in the shop, and be sure that the resistor is well-protected; it'll get pretty hot.

I'm restoring an old Philco 39-45 radio. It's got an odd 1st i-f transformer, which is open. It has 5 leads instead of four. Can I get a replacement for this?—E. N., Princeton, IL.

Not an exact replacement, but one that will work. This has a little tertiary winding, which is connected to the suppressor grid of the 78 i-f amplifier tube. The diagram (Fig. 9-13) shows how, but not why. (Actually, I don't know *why*. It may have been to neutralize a "hot" stage or to provide regeneration to hop up the stage for more gain.)

Something like a J. W. Miller 512-C1 input i-f transformer will work nicely. Just tie the suppressor grid of the 78 tube to its cathode, tune the i-f transformer to 470 kHz, (it is a 455 kHz type, but will easily tune to 470 kHz), and away you go.

I've built one of the burglar alarm circuits published in the June 1971 issue of Radio-Electronics. Works fine, but it will "false-trigger." Finally found that turning on the fluorescent lamp in the kitchen caused this. What can I do to stop this?—W. S., Buffalo, NY.

It sounds as if your alarm wiring is picking up some of the radio-frequency hash generated by some older fluorescent lamps. They can cause quite a lot, if unfiltered.

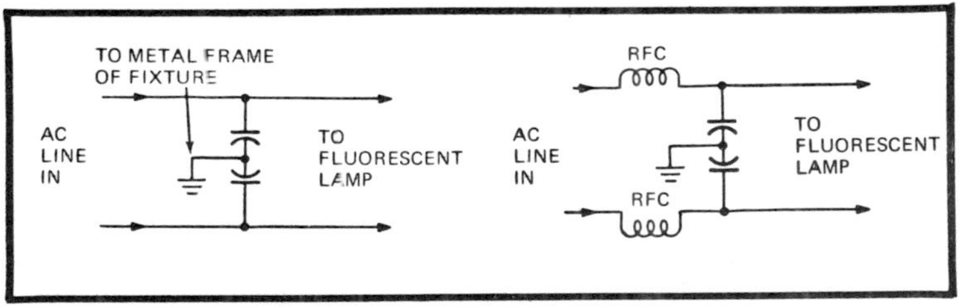

Fig. 9-14. Simple filters for eliminating fluorescent hash.

Two possibles. Shield the alarm wiring where it runs near the lamp wiring. One, add rf filters to the lamp circuit. Start with a couple of small bypass capacitors across the line (Fig. 9-14A). If this doesn't get rid of it, add a couple of small rf chokes as in diagram B. This ought to do it.

After I finally cleared up all of the shorts in this old Hallicrafters S-52 multiband receiver, I got a signal through the audio. However, I can't get an i-f or rf signal through from anywhere. Tubes good, dc voltages look fairly close.

I get about 75V on the plates and screens of the i-f and rf tubes, grids about 1.5V, and cathodes about 24—26V. This has an odd cathode circuit. Any ideas?—G. M., Greensboro, NC.

I've been stuck on the same thing. Simple, if you know it's there; tough if you don't. It's not a "stock" circuit. The diagram (Fig. 9-15) shows the cathode circuit. There's a 10,000-ohm SENSITIVITY control in series with the 6SG7 rf amplifier, as well as the two 6SK7 i-f amplifier cathodes. By changing the setting of this, you vary the *cathode bias* on all three tubes, thus changing the sensitivity.

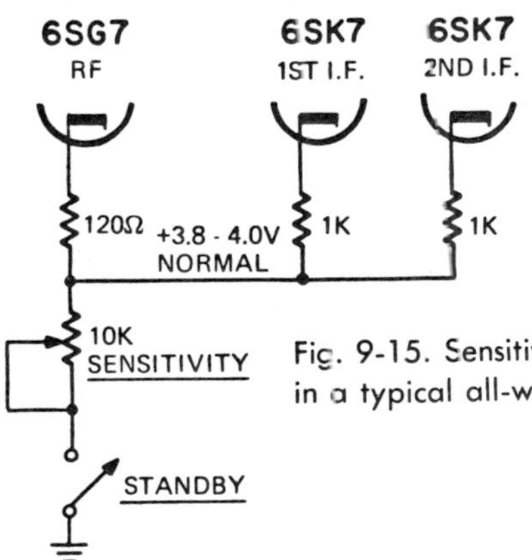

Fig. 9-15. Sensitivity or rf gain control circuit in a typical all-wave receiver.

If you'll notice the high cathode voltages you found, you'll see that these are high enough to cut the rf and i-f tubes completely off. Normal dc voltages here should be 3.8 to 4.0 volts. (Positive voltage on a cathode is the same as negative voltage on a grid.)

Try adjusting the SENSITIVITY control to see if you can get the bias back to normal. If it won't vary at all, the STANDBY switch in series with the ground return of the SENSITIVITY control is probably open. This cuts off the rf stage and i-f when you're transmitting.

I'm trying to fix a Kraco stereo tape player. Do you have a schematic for this, or know where I can get one? It has two blown 2SB487 transistors (power outputs) and I can't find them listed either! What is the replacement?—T. K., OH.

Nothing at all in my files on *Kraco*. You have an orphan. However, I did eventually locate a listing on a 2SB481 transistor, which is the same as the 2SB487. Motorola HEP-642, RCA SK-3052, GE-30, or Sylvania ECG-131. Be sure to check the little emitter resistors, which will probably be about 1.0 ohm each. If the original transistors shorted, these probably burnt up.

I replaced the stock antenna on a car with a 60-inch whip, rear-mounted, with RG-58/U coax. I've lost a lot of AM sensitivity, and the FM too is bad. Is this due to a mismatch between the 50-ohm RG-58/U coax and the input, or what, and what can I do about it?—B. D., Washington, D.C.

Yes it is. The only way they can make a car-radio work at all on that miserable little 3-foot rod is to design the *whole front end* of the radio to work with antennas within a certain very narrow range of capacitance and inductance. In other words, the whole thing, including the antenna, is *tuned*.

We used to have this same problem when the first rear-mount antennas were used. They used special lead-in cables, with a very small series capacitance.

Add a very small capacitor in series, say about a 10 pF. Then check your antenna trimmer on the radio. If it shows a peak, it'll work. If not, try a little different size.

This auto tape player will not change tracks. Even with the panel pushbutton, nothing happens.—W. H., Dunlap, IO.

Most of these use a soldenoid to move the head for track changing. Check the dc voltage across the solenoid terminals while holding the TRACK CHANGE button down. If you get voltage, disconnect the solenoid and check it for continuity. In several of these, you'll find a diode shunted across the coil for transient suppression. If it shorts, the solenoid won't work.

Index

A

Admiral
- 5H10 107
- 5H10 power supply 108
- 4G13 137
- 16M3 106
- 1H5 153, 220
- H10 187
- H3 219
- 3G11 144
- 2H5 153, 228
- H12 187
- 25E6 40
- G4 chassis 142

G13 187
- L33L31 107

Afc
- action erratic 231
- diodes bad 69

Afpc
- circuit requirements 27
- circuit & waveforms 27
- quick-checks 28
- troubles 26

Agc 149
- bypass open 150
- compound 153
- delay 148
- gate 154
- keying pulse 163
- problems 110, 146, 153, 163
- rf 165
- signal flow 154
- stage oscillation 171
- too high 164
- trouble 167
- won't work 228

Airline GEM-17148B 185

Amplitude
- tilt control 22

vertical sync 57

Ammeter/wattmeter 200

Anode voltage insufficient 221

Antenna preamp 156

Arcing
- focus rectifier 144
- in pix tube 145

Audio
- amplifier problems 236
- transistors blow 111
- volume too low 142
- xstr overheats 236

Automotive tests 204

B

B+
- absence of 142
- low 218
- rectifier burnt out 215
- too low 216

Bad
- bandpass bypass 20
- channel selector 40
- crystal 40
- flesh tones 38
- glass diode 111
- BPO capacitor 30
- purity 35
- solder joint 179
- sync 166
- thermal switch 107
- 3A3 socket 143
- ultor contact 145
- yoke socket 73

Bands
- of ripples 32, 163
- of streaks 160
- distortion 194

Bandpass bypass bad 20

Bars
- color 190
- horizontal 141
- vertical 43

Bending (S-curve) 183

Beta variations, xstr 98

Bias
- diode reversed 108
- problems 118
- problem, crt 145
- voltages 181

Black
- bars 131
- streaks 194

Blanked color 34
Blanking diode problems 131
Blink, vertical 185
Blooming 183, 224
  color 34
  in red 135
Blur across screen 184
Boosted-boost problems 132
Booster
  checking 157
  TV 156
Bowed lines 23, 36
Breaks in cables 199
Breaker pops with good diode 107
Bright line across screen 170
Brightness
  affects raster 220
  affects streaking 143
  causes bloom 224
  control causes streaking 216
  control, no 145
  control problem 141
  intermittent 185
  problems 220
Broken convergence wires 59
Burglar alarm false triggers 240
Burnt
  B + rectifier 215
  flyback 137, 225
  linearity control 70
  pincushion corrector plate resistor 109, 115, 211
  scope plate load 211
  VOM resistor 212
Burned
  cathode resistor 101
  degauss resistors 62
  focus socket 117
Burst
  amp problems 26
  checking for 27
Bypass open, agc 150

C

Cable
  breaks 199
  systems 188
  testing 204
Capacitance checking 198
Capacitor
  charge curve 53
  leaky 13, 44
  NPO 30
  open, filter 217
  short 140
  thermal 227
Catalina 122-770B 116
Cathode
  current climbs 44, 133
  heater shorts 120
  resistor burns 101
  voltage too high 62
  voltages wrong 141
CATV 188
  color loss 189
  crosstalk 190
  distortion 191
  ghost 189
Checking IC chips 80
Circuit breaker pops 103
Clairtone C11 183
Clamp diode bad 36
Coax cable mismatch 242
Coil, reactance 13
Collapsing horizontal scan 217
Colors
  absence of 16
  alignment 36
  bars 190
  bars, floating 41
  blobs 41
  blooming 34
  demod IC 82
  diodes unbalanced 29
  flicker 40
  fringing 16
  ghosts 20
  in & out 33
  intermittent 32, 228
  lock, poor 30
  missing 41
  oscillator 11
  out of phase 40
  problems 11
  purity loss 39
  rainbows 41
  reversed 31
  ringing 21
  slow 39
  smeared 19, 151
  sync bad 37, 41
  sync "loose" 30
  sync loss 31
  sync weak 28
  tests 29
  unparallel 23
  weak/vertical bars 43
  wrong 16
Comb patterns 24
Complement
  vertical circuits 59
  symmetry vertical sweep 59
Component problems 78
Contact problems, ultor 145
Contrast
  affects sync 230
  brightness affect raster 220

Controls, height/size 54
Convergence
controls 22
drift 42
dynamic 22
intermittent 33
less 31
troubles 36
wire broken 59
yoke socket 73
Corona discharge 160
Cracked pix tube 113
Craig 2108 tape recorder 239
Creeping cathode current 44
Critical hold control 228
Crosstalk, CATV 190
Crt
arcing 145
heater dead 131
heater out 136
lines 170
tracked 113
troubles 118
voltage too low 221
Crystal
demod probe 24
problems 40
CTC 17X 131

D

Damage by lightning 35
Damper
cathode, no volts 223
plate with no voltage 142
socket problem 139
Deflection
scope 206
yoke problem 77
Defocusing 123
Delay, agc 148
Demod
bypass open 20
defective 38
out of balance 17
Detector, killer 30
Detuned afc 228
Dim raster 134
Diode
bad 111
color tests 29
dropping 84
reversed 108
switching 97
tube replacement 91
varactor 93
DIP indexing 89
Disappearing raster 185
Discolored pix 38
Dissipation, power 78
Distortion
banding 194
CATV 191
in 1 stereo channel 236
stereo 183
Dots and streaks 160
Drift
convergence 42
VTVM 210
Drifting scope calibration 206
Drive
lines 135
line in raster 63
Dropping diode 84
Dynamic convergence 22

E

Electrical interference 161
Emerson
120771 64, 65, 226
129021 131
835A 77
Emission, grid 18, 45
Error, voltage 193
Excess
agc voltage 166
cathode current 110
color blanking 34
snow 162
vertical retrace 70

F

Faces too red 38
Fading 183
False
keystone 69
triggering burglar alarm 240
Feedback, vertical 55
Filter
bad 172
open 217
Finding cable breaks 199
Flare, intermittent 183
Flashes 184
& streaks 145
Flashing, focus rectifier 144
Flesh tone bad 38
Flickering colors 40, 164
Floating
bars in color 41
blur 184
Flyback
bad 163
burns out 225
burnt out 137
replacement 111
resistance problem 110
transformer 125
troubles 107, 218
xstr shorts 127
FM
dead, AM OK 237
receiver problems 105

Focus 122
current 123
intermittent 184
problem 116
pulses 139
rectifier arcing 144
rectifier problem 124
socket corroded 134
transformer 117
trouble 129, 135
voltage, absence of 117
voltage, too low 222
Foldover 103
vertical 108
Four black bars 131
Frequency response low 195
Fringing, color 16
Frying sound in flyback 63
Fuse-blowing 35, 220, 225

G

Gain chart 158
Garbled channel, tape recorder 223
Gate
leaky 149
turnoff problem 64
turnoff switch 226
GE
AA chassis 183
CA chassis 217
CB 133
G13 40
H-3 34
JA vertical ckt 60
MA vertical size 62
M110YBG 104
M587 138
M760CWD 77
SB chassis 228
14P1215 103
Generator, sweep (low output) 212
Ghosts
in color 20
on cable 189
Glitch on waveform 91
Glow on screen 141
Good diode blows breaker 107
Greens, excessive 39
Grid
emission 45
resistor heats 220
waveforms, horizontal 45
GTO switch 226

H

Halved pix 61
Half-width raster 216
Hallicrafters S-52 241
Heater
cathode shorts 120
out, crt 136
Heath
GR-25 67
GR-295 132, 222
IG-14 214
IM-13 VTVM 210
10-12 scope 206
Heat
dissipation, transistor 78
up distortion 183
Height
loss 58, 73
problems 62
vertical size 54
High-pitched whistle 76
Horizontal
blanker & bias 119
blanking bar locks 226
blanker waveforms 43
coil has no effect 104
drift 69
driver blows 66
hold coil loose 227
hold critical 228
hold range insufficient 63
hold range shrunk 226
hold unstable 230
instability 63
jitter 227
line 170
line rippling 67
oscillator out 219
output waveforms 45
pulling 151
scan collapses 217
scan lines 123
shrinking 64
smearing 143, 162
stabilizer coil problem 104
streaking 145
sweep, absence of 77
sync dropout 186
sync gone 104
sync problems 165
sync out 109
vertical sync problems 231
troubles 70
Hot chassis 220
Hum
bars in color 38
on intercom 223
HV
absence of 218, 219
and sweep missing 61
cage smoking 116
problems 142
protect circuit 221
rectifier 92

rectifier overload 225
regulator 223
too low 216
Hybrid video problems 18

I

IC
pin numbering 88
removal 101
voltage checks 87
Imbalanced demod 17
Impedance matching problem 232
Indicator-lamp failure 115
Ineffective agc 228
Insufficient
audio 142
brightness control 145
B + 216
crt anode voltage 221
focus voltage 222
height 47, 62
horizontal hold range 226
HV 142
red 38
screen control 145
signal 162
video 162
width 77
Intercom hum 223
Interference 111, 160
Intermittents 168
brightness 185
color 15, 33, 35, 228
convergence 33
flare 133
focus problem 124
short 187
smearing 151
streaking 145
sync 186
VTVM 213
Isolating the color oscillator 28

J

Jig, smear 196
Jitter
horizontal 227
vertical 90
with sound 227
Junction leakage, xstr 46

K

Keystone, false 69
Keying
pulse, agc 163
pulse phase 15
Killer detector 30
Knight
KG-630 scope 213
KG-635 scope 211
tape recorder 233
Kraco 242

L

Lafayette RK890 A 8-track 232
Leakage, transistor 46
Leaky
afc diode 230
capacitor 138
capacitor (vertical) 57
coupling capacitor 44
gate 149
gate turnoff 64
Lightning damage 35, 220
Linearity, vertical 54
Line
bowed 36
drive 135
on screen 170
retrace 183
squeezing, scan 64
Load loops 125
Local oscillator 12
Locating rfi sources 161
Lock, color (bad) 30
Locked
blanking bar 63
horizontal blanking bar 226
Loops, load 125
Loose
color sync 30
horizontal hold coil 227
Loss of
CATV color 189
convergence 31
height 73
pilot carrier 238
purity 39
raster 118
of eds 38
of sync 151
Low
crt anode voltage 221
focus voltage 222
high voltage 216
level recording on 1 track 232
sweep generator output 212
video 162

M

Magnavox
CR-703 235
1T-109 186
T-924 41
904 134
908 48, 70, 150
919 143
920 127
T933 41, 89
938 22

Many pix, 1 screen 73
MATV
  line amplifier 158
  problems 196
Mayfair FT-1024B 105
Meter shunts 210
Middle-stretch 77
Midland 23-101 VOM 212
Mixer noise 148
Motorboating xstr radio 233
Motorola 8X26E 233
Motorola
  GP20GW cassette deck 232
  TS-454YC 165
  TS-544 162
  551 104
  TS-579-A 216
  TS-597 109, 231
  TS-914 39, 42, 221
  TS-918 31
  TS-929 32, 163
  TS-934A 231
Moving color bars 41
MS Hammond M organ 126
Multivibrator 50
Mute switch malfunction 105

N

Narrow raster 226
Negative pix 19
NE-2 replacement 115
No
  B+ 142
  B+ boost 132
  brightness control 145
  color 16, 41
  color sync 31, 37
  control of bolume 103
  deflection on scope 75
  focus 135
  focus voltage 117, 222
  high voltage 218, 219
  horizontal oscillation 219
  horizontal sweep 77
  horizontal sync 104
  multiplex 105
  pedal tones (organ) 236
  pix tube heater 136
  raster 118, 133, 138
  red setup line 33
  sawtooth 65
  scope position control 207
  sync, B & W 229
  vertical deflection, scope 213
  vertical sweep 65, 73, 170, 174
  voltage okay 134
Noise 146, 156
  mixer 148

O

Odd
  colors 16, 32
  raster 141
Off-frequency 3.58 MHz oscillator 14
Olympic
  CT-910 137
  3P48 112
Oversensitive burglar alarm 240
Open
  agc bypass 150
  bandpass resistor 21
  i-f xfmr (radio) 240
  filter capacitor 217
One-channel color problems 40
Oscillation 171
Oscillator
  color 11
  failure (3.58) 13
  local 12
  testing 14
  vertical 51
Output low, sweep generator 212
Overheating grid resistor 220
Overloaded HV rectifier 225
Oversaturation 40
Oxidation, tube pin 140

P

Panasonic
  AN-32 69
  8-track 232
PC
  board shorts 19
  connection bad 179
  shorts 186
Phase control 12
Philco
  12N50A 230
  18QT85 129
  19KT51 184
  39—45 radio 240
  Q1054BK 66
Picture
  chopped at top and bottom 227
  color loss 35
  disappears 131, 164
  drift, horizontal 69
  half missing 61
  smearing 162
  too small 41, 47
  tube arcing 145
  tube bias problem 118
  tube cracked 113

tube heater dead 131
tube heater out 136
tube voltage low 221
won't fill screen 227
Pilot carrier loss (FM) 238
Pin correction trouble 67
Pincushion trouble 68
Pioneer
Series 8 tape player 102
SX-1500T stereo 105
Plates turn red, rectifier 222
Plate-to-heater arcing 144
Player won't switch 242
Plug-in semiconductor diode 91
Plymouth 9BBCC radio 238
Pocket joint 179
Power
dissipation 78
supply diodes reversed 235
Preamps 156
Precision E-200 generator 208
Prescaler 202
Pulling horizontal raster 133, 151
Pulse, agc keying 163
Pulsing focus 139
Purity
loss 39
problems 38, 42

Q

Quasar II 32, 163

R

Radio distorts after warmup 234
Rainbows
in color 41
in picture 27
Raster
absence of 118, 133, 138
blooms 224
collapse 183
dim 134
disappears 185
drive line 63
drops out 220
half-width 216
narrow 226
odd 141
pulling 138
streaks 143
stretch 77
trapezoidal 137
RCA
CTC 9 184
CTC 11 38
CTC 15 185
CTC 16 39
CTC 17X 110, 145
CTC 19 33
CTC 22 128
CTC 24 63, 218
CTC 25 216
CTC 28 43, 134
CTC 31 175
CTC 38 136, 151
CTC 38XL 145
CTC 40 106
CTC 44B 41
CTC 55 75
CTC 63XC 220
HSK-T1 103
KCS-130 73
KCS-136 166
KCS-136M 70
KCS-144 230
KCS-153 74
KCS-160 114
KCS-169 46
KCS-176A 84
Synchroguide 74
TVI filter 112
211T1-T3 141
WO-33A 206
Reactance
coil 13
tube 12
Rectifier
arcing, focus 144
gets red hot 222
overload (HV) 225
reversed, stereo 235
Red
bloom 135
and blue missing 39
flicker 40
hot rectifier plates 222
los of 38
setup line gone 33
Regulator tube draws excess current 222
Regulated voltage off 218
Regulator
circuit 223
doesn't regulate 222
Remote control on scope 107
Removing ICs from board 101
Resistor
burnout 109
burnt, VOM 212
(grid) heats 220
voltage-dependent 55
Retrace lines 183
Reversed
bias diode 108

B+ rectifier 218
colors 31
Rf
agc problems 165
output msmt 202
Ringing
color 21
in yoke 136
1X2 plate 47
Ripple
bands 163
in horizontal lines 67
Ripples 32
Rolling, intermittent 184

S

Sawtooth
absence of 65
from scope 206
S-bending 183
Scan lines squeezed 64
Scope
calibrated drift 206
prescaler 202
problems 211
sawtooth 206
vertical position trouble 207
without vertical deflection 213
xfmr shorted 210
Screen
flashes 18
tinted 15
Sears
528.81201 133
9168 164
Service switch affects
scan 217
Setchell-Carlson U802 115
Setup line missing 33
Sharp CN-52T 185
Sherwood S-7100A stereo 238
Short, intermittent 187
Shorted
capacitor 140
open flyback 218
PC board 19
scope xfmr 210
transistor 19
Shorts
flyback xfmr 127
heater-cathode 120
PC 186
Shrink, thermal 226
Shunt, meter 210
Signal
flow, agc 154
generator 208
Silvertone 8175 130
Singer TV6 227
Skewing, videotape 191
Slow
color warmup 39
warmup 162
Smeared raster 116
jig 196
Smearing 162
Smeary color 19, 41, 151
Smoke from HV cage 116
Snow 162
Socket problem, 3A3 143
Sony
9-304UW 225
TC-102 tape deck 236
Sound clear but low 142
Sparkle flecks 18
Spikes, vertical frequency 65
Split picture 228
Squeezed scan lines 64
SSHVR 92
Stereo
blows fuses 234
distortion 183
Streaking
black 194
with brightness control
variation 216
Streaks
and dots 160
and flashes 145
in raster 143
Stretch in raster 77
Stromberg-Carlson TC-19 141
Sweep
absence of 170
and deflection 44
generator output low 212
horizontal (none) 77
low partial 59
vertical (complementary) 59
vertical (no) 174
Switching
diode 97
transients 42
Synchroguide 74
Sylvania
D-12 37, 218
D14 186
D-15 31
DO-3 170
DO-9 135
132 scope 75, 213
580-1 220
factory modification 186
Sync
absence of 41
amplitude, vertical 57
and audio detector 167
bad 166, 230, 231

color (bad) 37
horizontal (no) 104
instability 183
intermittent 186
loss 151
loss of (vertical) 57
problems 229
problems, horizontal 109, 165
vertical 56
videotape 197
weak, color 28

T

Tape
okay on playback but distorts on record 239
player audio 105
player won't track-change 242
Tearing raster 138
Test
equipment 88, 198
instruments 126
setup, rf booster 157
Tester, flyback 126
Testing the vdr 56
Thermal
capacitor 227
resistance 79
shrink 226
switch bad 187
Thermistor biasing 105
3.58 MHz
oscillator 11
oscillator tests 14
Tilt/amplitude control 22
Tinted screen 15
Track change won't work 242
Transformer, flyback 125
Transformerless vertical output 59
Transistor
burn out 105
dissipation 78
maximum power 78
radio motorboats 233
ratings 98
shorted 19
thermal resistance 79
Trap
doesn't stop TVI 111
setting, markers 214
setting problems 72
Trapezoid raster 137
Troubleshooting 169, 173
Trouble tree 173
Truetone
4DC5929A-86 stereo 236
EIS2885A-86 141
MIC3919C 111
stereo output circuit 237
Tube pin oxidation 140
Tuner
off channel 164
problems 146, 164
Tuning diode 93
Turnoff switch, gate 226
TV boosters 156
TVI
trap 112
with trap 111

U

Ultor contact bad 145
Unbalanced diodes (color) 29
Undamped bandpass xfmr 21
Unparallel colors 23
Unstable hold 230

W

Warmup too slow 162
Wards GHJ-949B stereo 104
Washed-out picture 166
Wattmeter, homebrew 200
Waveform
glitch 91
horiz output 45
and traps 72
Weak
color sync 28
sync (vertical) 57
video 130
Westinghouse
BP19A770 103
V2486-2 108
Whine from set 76
Whistle in audio 76
White
flashes 184
vertical line 52, 74
Window-shade effect 141
Wrong
cathode voltages 141
color 15

Y

Y-channel problems 18
Yoke
balance 73
ringing 136
Yokes 51

Z

Zenith
9ZT15 radio 237
10YT26 stereo 234

12B8C15 33
14A9C19 38
14A9C50 38, 39, 40
14N22 90, 230
14N23 218
14N28 184
14M29 76
14Y33 64
14Z8C50 82
15L33 109
16D21Q 165
16Z7C17 31
16Z7C50Z 164
19CC19 224
19DC20 111
20AT30Z stereo 111
20X1C38 32, 66
20Y1C38 34, 35, 70
20Y1C48 62, 131
20Y1C50 69, 225
20X1C36 228
22ABS5 229
23XC36 215
23XC38 42
24NC31Z 62, 73
25LC30 117
25MC33 63, 139, 226
25MC36 227
25NC38 222
25NC38 regulator 223
Zero volts on
damper cathode 223